U.S. Marine Corps Close Combat

Plus

Marine Physical Fitness Training and Test

Department of Defense

Pentagon
Publishing

DEPARTMENT OF THE NAVY
Headquarters United States Marine Corps
Washington, D.C. 20380-1775

18 February 1999

FOREWORD

1. PURPOSE

Today's Marines operate within a continuum of force where conflict may change from low intensity to high intensity over a matter of hours. Marines are also engaged in many military operations other than war, such as peacekeeping missions or noncombatant evacuation operations, where deadly force may not be authorized. During noncombative engagements, Marines must determine if a situation warrants applying deadly force. Sometimes Marines must decide in a matter of seconds because their lives or the lives of others depend on their actions. To make the right decision, Marines must understand both the lethal and nonlethal close combat techniques needed to handle the situation responsibly without escalating the violence unnecessarily. Marine Corps Reference Publication (MCRP) 3-02B, *Close Combat,* provides the tactics, techniques, and procedures of Marine Corps close combat. It also provides the doctrinal basis for the Marine Corps Close Combat Training Program (MCCCTP).

2. SCOPE

This publication guides individual Marines, unit leaders, and close combat instructors in the proper tactics, techniques, and procedures for close combat. MCRP 3-02B is not intended to replace supervision by appropriate unit leaders and close combat instruction by qualified instructors. Its role is to ensure standardization and execution of tactics, techniques, and procedures throughout the Marine Corps.

3. SUPERSESSION

MCRP 3-02B supersedes Fleet Marine Force Manual (FMFM) 0-7, *Close Combat,* dated 9 July 1993. There are significant differences between the two publications. MCRP 3-02B should be reviewed in its entirety.

4. WARNING

Techniques described in this publication can cause serious injury or death. Practical application in the training of these techniques will be conducted in strict accordance with approved Entry Level Close Combat, Close Combat Instructor (CCI), and Close Combat Instructor Trainer (CCIT) lesson plans. Where serious danger exists, the reac er is alerted by the following:

WARNING

5. CERTIFICATION

Reviewed and approved this date.

BY DIRECTION OF THE COMMANDANT OF THE MARINE CORPS

J. E. RHODES
Lieutenant General, U.S. Marine Corps
Commanding General
Marine Corps Combat Development Command

DISTRIBUTION: 14400006600

OVERVIEW OF CLOSE COMBAT

1. Purpose of Close Combat

Close combat is the physical confrontation between two or more opponents. It involves armed and unarmed and lethal and nonlethal fighting techniques that range from enforced compliance to deadly force. The purpose of close combat is to execute armed and unarmed techniques to produce both lethal and nonlethal results. Unarmed techniques include hand-to-hand combat and defense against hand-held weapons. Armed techniques include techniques applied with a rifle, bayonet, knife, baton, or any weapon of opportunity.

2. Continuum of Force

Marines will find themselves in both combative and noncombative situations. The threat level in these situations can rise and fall several times based on the actions of both Marines and the people involved. The escalation of force stops when one person complies with the demands imposed by another person. This range of actions is known as a continuum of force. Continuum of force is the concept that there is a wide range of possible actions, ranging from voice commands to application of deadly force, that may be used to gain and maintain control of a potentially dangerous situation (MCO 5500.6_, *Arming of Security and Law Enforcement [EE] Personnel and the Use of Force).* Continuum of force consists of five levels that correspond to the behavior of the people involved and the actions Marines use to handle the situation (see the table below). Close combat techniques are executed in levels three, four, and five.

Level One: Compliant(Cooperative)

The subject complies with verbal commands. Close combat techniques do not apply.

Level Two: Resistant (Passive)

The subject resists verbal commands but complies immediately to any contact controls. Close combat techniques do not apply.

Continuum of Force		
Level	**Description**	**Actions**
1	Compliant (Cooperative)	Verbal commands
2	Resistant (Passive)	Contact controls
3	Resistant (Active)	Compliance techniques
4	Assaultive (Bodily Harm)	Defensive tactics
5	Assaultive (Serious Bodily Harm/ Death)	Deadly force
Note: Shading indicates levels in which Marines use close combat techniques.		

Level Three: Resistant (Active)

The subject initially demonstrates physical resistance. Marines use compliance techniques to control the situation. Level three incorporates close combat techniques to physically force a subject to comply. Techniques include—

- Come-along holds.
- Soft-handed stunning blows.
- Pain compliance through joint manipulation and the use of pressure points.

Level Four: Assaultive (Bodily Harm)

The subject may physically attack Marines, but he does not use a weapon. Marines use defensive tactics to neutralize the threat. Defensive tactics include the following close combat techniques:

- Blocks.
- Strikes.
- Kicks.
- Enhanced pain compliance procedures.
- Nightstick blocks and blows.

Level Five: Assaultive (Serious Bodily Harm/Death)

The subject usually has a weapon and will either kill or seriously injure someone if he is not stopped immediately and brought under control. Typically, to control the subject, Marines apply deadly force through the use of a firearm, but they may also use armed and unarmed close combat techniques.

3. Marine Corps Tactical Concepts

Close combat techniques support the following key Marine Corps tactical concepts. The concepts are not standalone ideas but are to be combined to achieve an effect that is greater than their separate sum.

Achieving a Decision

Achieving a decision is important in close combat. An indecisive fight wastes energy and possibly Marines' lives. Whether the intent is to control an opponent through restraint or defend themselves in war, Marines must have a clear purpose before engaging in close combat and act decisively once engaged.

Gaining an Advantage

A basic principle of martial arts is to use the opponent's strength and momentum against him to gain more leverage than one's own muscles alone can generate, thereby gaining an advantage. In close combat, Marines must exploit every advantage over an opponent to ensure a successful outcome. This can include employing various weapons and close combat techniques that will present a dilemma to an opponent. Achieving surprise can also greatly increase leverage. Marines try to achieve surprise through deception, stealth, and ambiguity.

Speed

Marines use speed to gain the initiative and advantage over the enemy. In close combat, the speed and violence of the attack against an opponent provides Marines with a distinct advantage. Marines must know and understand the basics of close combat so they can act instinctively with speed to execute close combat techniques.

Adapting

Close combat can be characterized by friction, uncertainty, disorder, and rapid change. Each situation is a unique combination of shifting factors that cannot be controlled with precision or certainty. For example, a crowd control mission may call for Marines to employ various techniques ranging from nonlethal restraint to more forceful applications. Marines who adapt quickly will have a significant advantage.

Exploiting Success

Typically, an enemy will not normally surrender simply because he was placed at a disadvantage. Marines cannot be satisfied with gaining an advantage in a close combat situation. They must exploit any advantage aggressively and ruthlessly until an opportunity arises to completely dominate the opponent. Marines must exploit success by using every advantage that can be gained.

(reverse blank)

Close Combat

Table of Contents

CHAPTER 1

FUNDAMENTALS OF CLOSE COMBAT

This chapter describes all techniques for a right-handed person. However, all techniques can be executed from eittwr side.

The Marine is depicted in camouflage utilities. The opponent is depicted without camouflage.

The fundamentals of close combat include ranges, weapons of the body, target areas of the body, and pressure points of the body. These fundamentals form the basis for all close combat techniques. They provide Marines with a common framework regardless of the type of confrontation or the techniques used. If Marines apply these fundamentals properly in a close combat situation, they may save their lives or the lives of fellow Marines.

1. Ranges of Close Combat

Close combat engagements occur within three ranges: long range, midrange, and close range. During any engagement, these ranges may blur together or may rapidly transition from one to another until either the opponent is defeated or the conflict is resolved.

Long Range

During long range engagements, combatants engage each other with rifles, bayonets, sticks, or entrenching tools. See figure below.

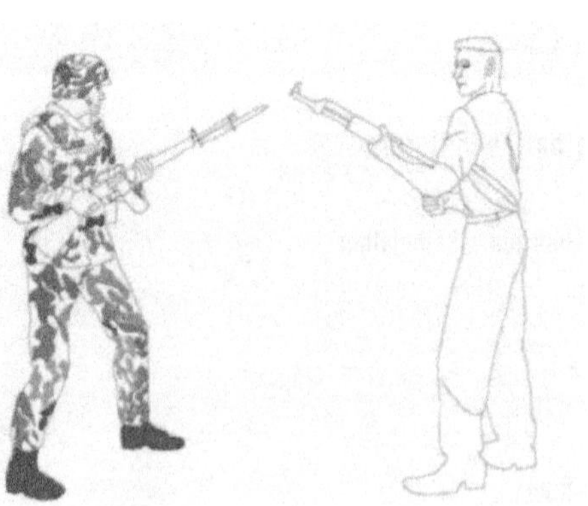

Midrange

During midrange engagements, combatants engage each other with knives, punches, or kicks.

Close Range

During close range engagements, combatants grab each other. Close range engagements also involve elbow strikes, knee strikes, and grappling.

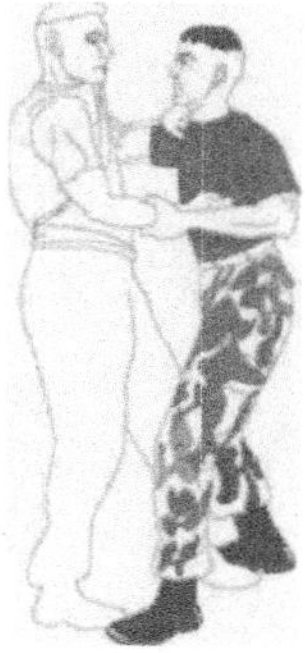

2. Weapons of the Body

Hands and Arms

The hands, forearms, and elbows are the arm's individual weapons. The hands consist of several areas that can be used as weapons: fists, edges of hands, palms, and fingers.

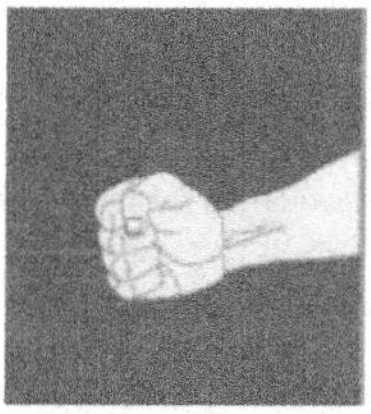

Fists. To minimize injury to the fists, Marines use their fists as weapons to target soft tissue areas such as the throat. The fists' striking surfaces are the first two knuckles of the hands or the meaty portions of the hands below the little fingers.

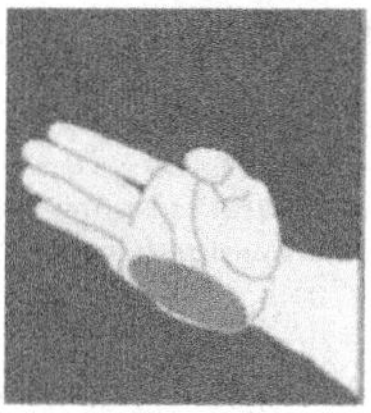

Edge of Hand. Marines use the edge of the hand (knife edge) as a weapon. Marines use the edge of the hand to strike soft tissue areas.

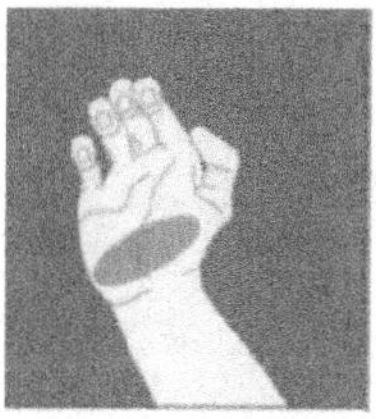

Palms. Because of the palm's padding, Marines use the heels of the palms to strike, parry, and/or block.

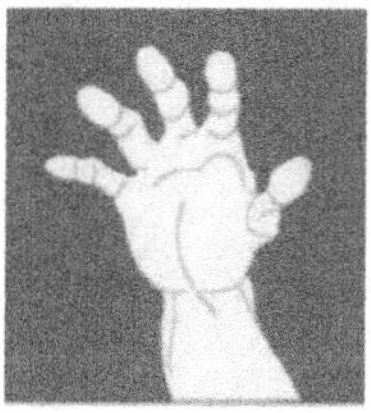

Fingers. Marines use the fingers to gouge, rip, and tear soft tissue areas (e.g., eyes, throat, groin).

Forearms. Marines use the forearms as a defensive tool to deflect or block attacks. Forearms can also be used as striking weapons to damage or break an opponent's joints and limbs. Marines sustain less self-injury when strikes are conducted with the forearms than when strikes are conducted with fists and fingers.

Elbows. Marines use the elbows as striking weapons. Because of the short distance needed to generate power, elbows are excellent weapons for striking during the close range of close combat.

Legs

The legs are more powerful than any other weapon of the body, and they are less prone to injury when striking. The feet are protected by boots and are the preferred choice for striking.

Feet. Marines use the balls of the feet, the insteps, and the toes to kick an opponent. Marines use the cutting edge of the heels and the heels to stomp on an opponent. Marines must be wearing boots when striking with the toes.

Knees. Like elbows, knees are excellent weapons in the close range of close combat. Knee strikes are most effective while fighting close to an opponent where kicks arc impractical. The opponent's groin area is an ideal target for the knee strike if he is standing upright. Knee strikes can deliver a devastating secondary attack to an opponent's face following an initial attack that caused him to bend at the waist.

3. Target Areas of the Body

During close combat, Marines strive to attack the accessible target areas of an opponent's body. The readily accessible areas will vary with each situation and throughout the engagement. The target areas are divided into five major groups: head, neck, torso, groin, and extremities. The figure below illustrates target areas of the body.

Head

The vulnerable regions of the head are the eyes, temple, nose, ears, and jaw. Massive damage to the head kills an opponent.

Eyes. The eyes are excellent targets because they are soft tissue areas that arc not protected by bone or muscle. Attacks to this area may cause the opponent to protect the area with his hands, allowing Marines to execute a secondary attack to other

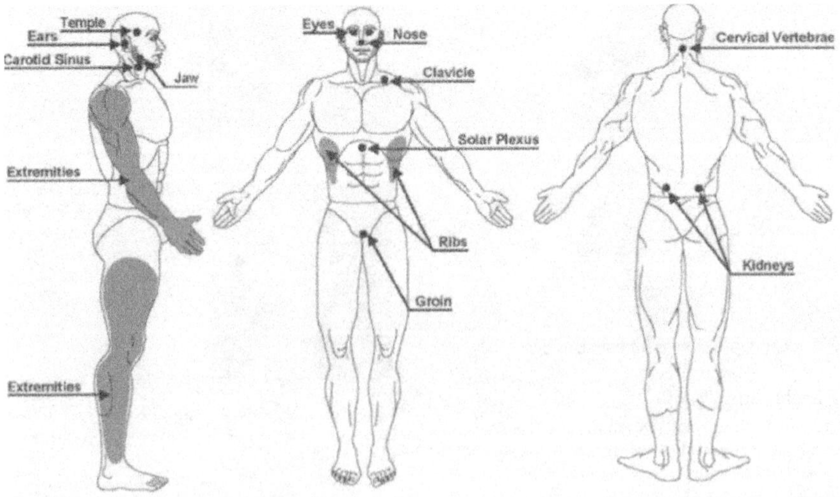

target areas while the opponent uses his hands to protect his eyes.

Temple. The temple is one of the most fragile areas of the head. Powerful strikes to the opponent's temple cause permanent damage and death.

Nose. The nose is very sensitive and easily broken. An attack to this area causes involuntary watering and closing of the opponent's eyes, rendering him vulnerable to secondary attacks. However, through training, individuals can condition themselves to withstand attacks to the nose. Therefore, any attack to the nose must be powerfully delivered.

Ears. Attacks to the ears may cause the eardrum to rupture. But this may not stop or even distract an opponent unless Marines powerfully deliver the strike.

Jaw. The jaw region, when struck forcefully, renders the opponent unconscious. Strikes to the jaw cause painful injuries to the teeth and surrounding tissues (e.g., lips, tongue), but the risk of self-injury is great unless Marines deliver strikes with a hard object such as a helmet, rifle butt, or boot heel.

Neck

The front of the neck, or throat area, is a soft tissue area that is not covered by natural protection. Damage to this region causes the opponent's trachea to swell, closing his airway, which can lead to death.

Carotid Sinus. The carotid sinus is located on both sides of the neck just below the jaw. Strikes to the carotid sinus restrict blood flow to the brain, causing loss of consciousness or death.

Cervical Vertebrae. The cervical vertebrae on the back of the neck, from the base of the skull to the top of the shoulders, contains the spinal cord, which is the nervous system's link to the brain. The weight of the head and the lack of large muscle mass allow damage to the cervical vertebrae and spinal cord. Excessive damage to this area causes pain, paralysis, or death.

Torso

Clavicle. The opponent's clavicle (or collar bone) can be easily fractured, causing immobilization of the arm.

Solar Plexus. Attacks to the opponent's solar plexus or center of the chest can knock the breath out of him and immobilize him.

Ribs. Damage to the opponent's ribs immobilizes him. It may also cause internal trauma.

Kidneys. Powerful attacks to the opponent's kidneys cause immobilization, permanent damage, or death.

Groin

The groin area is another soft tissue area not covered by natural protection. Any damage to this area causes the opponent to involuntarily protect his injured area, usually with his hands or legs. In male opponents, the scrotum is the main target since even a near miss causes severe pain, contraction of the lower abdominal muscles, deterioration of his stance, and possible internal trauma.

Extremities

Rarely will an attack to the opponent's extremities (arms and legs) cause death, but they are important target areas in close combat. Damage to an opponent's joints causes immobilization.

4. Pressure Points of the Body

There are nerves in the human body that, when pressure is applied or when they are struck, allow Marines to control a subject through pain compliance. Marines use pressure points to control an opponent when deadly force is not authorized. They also use pressure points to soften or distract an opponent so a lethal or nonlethal technique can be employed. The figure on page 1-5 illustrates the body's pressure points. Marines execute attacks to pressure points by—

: Rapidly kicking or striking pressure points.
: Slowly applying steady pressure to pressure points.

Infraorbital Nerve

The infraorbital nerve is just below the nose. Marines apply pressure to this nerve with an index finger to control the opponent.

Mastoid Process

The mastoid process is behind the base of the ear and beneath the edge of the jaw. Marines apply inward and upward pressure to this pressure point with the fingers to distract and control the opponent.

Jugular Notch

The jugular notch is at the base of the neck in the notch formed at the center of the clavicle. Marines apply pressure in a quick, stabbing motion with the index finger. Strikes to the jugular notch cause serious damage.

Brachial Plexus (Tie In)

The brachial plexus (tie in) is on the front of the shoulder at the joint. Strikes and pressure applied with the hand are effective on this nerve.

Radial Nerves

Radial nerves are on the inside of the forearms along the radius bones. Strikes and pressure applied with the hand to the radial nerve serve as a softening technique.

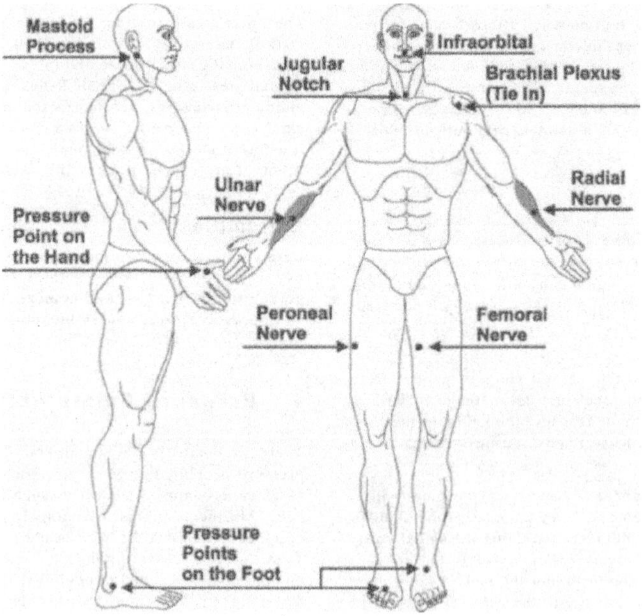

Ulnar Nerve

Ulnar nerves are on the outside of the forearms along the ulnar bones. Strikes and pressure applied with the hand to the ulnar nerve serve as a softening technique.

Pressure Point on the Hand

The hands contain a pressure point on the webbing between the thumbs and index fingers where the two bones of the fingers meet.

To force an opponent to soften or release his grip, Marines apply pressure with their index fingers to this pressure point or strike this pressure point with their fists.

Femoral Nerves

Femoral nerves are on the inside of the thighs along the femur bones. Strikes to the femoral nerve serve as a softening technique.

Peroneal Nerves

Peroneal nerves are on the outside of the thighs along the femur bones. Strikes to the peroneal nerve serve as a softening technique.

Pressure Points on the Feet

There are pressure points on the feet that, when pressure is applied or when they are struck, serve to soften or distract the opponent. Marines apply pressure with the toe, edge, or heel of their boots to the following points:

: The notch below the ball of the ankle.
: The top center of the foot, above the toes.
: The top of the foot where the leg and foot meet.

5. Basic Warrior Stance

Achieving a solid stance is critical to stability and movement throughout any close combat confrontation. The basic warrior stance provides the foundation for stability and movement that is needed to execute close combat techniques. To execute the basic warrior stance, Marines put their feet apart, hands up, elbows in, and chin down.

Feet Apart

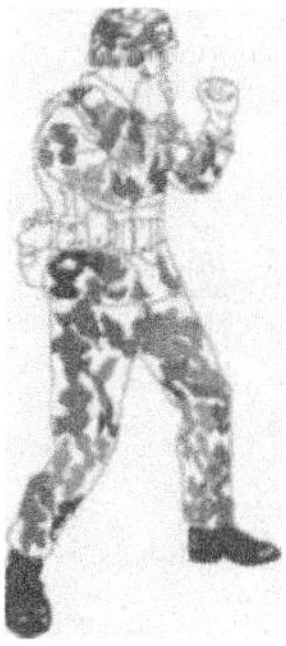

Place feet shoulder-width apart.

Keep the head forward and eyes on the opponent, take a half step forward with the left foot, and pivot on the heels so the hips and shoulders are at approximately a 45-degree angle to the right.

Distribute body weight evenly on both legs. Bend the knees slightly.

Hands Up

Curl the fingers naturally into the palm of the hand. Position the thumb across the index and middle fingers. Do not clench the fists. Clenching the fists increases muscular tension in the forearms and decreases speed and reaction time.

Bring the hands up to the face at chin level, with the palms facing each other. Hold the fists up high enough to protect the head, but not so high that they block the field of vision. Ensure continuous eye contact with the opponent.

Elbows In

Tuck the elbows in close to protect the body.

Chin Down

Tuck the chin down to take advantage of the natural protection provided by the shoulders.

6. Angles of Approach and Movement

Marines use movement to control a confrontation and to retain a tactical advantage. Movement increases power and maximizes momentum. By moving around the opponent, Marines gain access to different target areas of the opponent's body. Movement allows Marines to use different weapons of the body and different close combat techniques to attack specific target areas.

Angles of Approach

Marines move anywhere within a 360-degree circle around the opponent to gain a tactical advantage. This circle provides access to different target areas of the opponent's body.

When facing an opponent, Marines move in a 45-degree angle to either side of the opponent. Moving at a 45-degree angle avoids an opponent's strike and puts Marines in the best position to attack the opponent. Marines should avoid being directly in front of an opponent during a confrontation. If a Marine is directly in front of an opponent, the opponent can rely on his forward momentum and linear power to seize the tactical advantage.

Movement

Marines must know how to move in all directions while maintaining the basic warrior stance. During any movement, the legs or feet should not be crossed. Once a movement is completed, the basic warrior stance should be resumed. Maintaining the basic warrior stance protects Marines and puts them in the proper position to launch an attack against an opponent.

Note: Before body movement begins, Marines turn their heads quickly to the new direction. The faster the head turns, the faster the body moves, and the quicker Marines attain visual contact with the opponent.

Forward to the Left. To move forward to the left from the basic warrior stance, Marines-

: Move the left foot forward at a 45-degree angle from the body (approximately 12 to 15 inches), keeping the toe pointed toward the opponent.

Bring the right foot behind the left foot as soon as the left foot is in place. This returns Marines to the basic warrior stance.

Forward to the Right. To move forward to the right from the basic warrior stance, Marines-

Move the right foot forward at a 45-degree angle from the body (approximately 12 to 15 inches).

Bring the left foot, toe pointing toward the opponent, in front of the right foot as soon as the right foot is in place. This returns Marines to the basic warrior stance.

Backward to the Left. To move backward to the left from the basic warrior stance, Marines execute the forward movement in reverse. Marines-

Move the left foot backward at a 45-degree angle from the body (approximately 12 to 15 inches), keeping the toe pointed toward the opponent.

Bring the right foot behind the left foot as soon as the left foot is in place. This returns Marines to the basic warrior stance.

Backward to the Right. To move backward to the right from the basic warrior stance, Marines execute the forward movement in reverse. Marines-

Move the right foot backward at a 45-degree angle from the body (approximately 12 to 15 inches).

Bring the left foot, toe pointing toward the opponent, in front of the right foot as soon as the right foot is in place. This returns Marines to the basic warrior stance.

7. Balance and Off-Balancing

Balance

In any close combat situation, Marines must strive to maintain balance. The last place to be in a close combat situation is on the ground. Marines must maintain a strong base and a low center of balance, their feet must be a shoulder-width apart, and they must stay on their toes to enable quick movement.

Off-balancing

Marines use off-balancing techniques to control an opponent. These techniques are used to throw an opponent to the ground while Marines remain standing, or they are used to put Marines in a position for an offensive attack.

Off-balancing techniques use the opponent's momentum to move or throw him. For example, if the opponent is charging a Marine, the Marine pulls the opponent to drive him to the ground. Likewise, if the opponent is pulling a Marine, the Marine pushes the opponent to drive him to the ground.

Off-balancing techniques also rely on the power generated by the opponent. For example, during combat a Marine may be tired or outnumbered. Depending on the generated energy and momentum of the opponent, the Marine employs off-balancing techniques with very little effort and still provides effective results.

Because off-balancing techniques rely on the momentum and power generated by the opponent, these techniques are particularly effective for Marines who may be outsized by their opponent or lack their opponent's strength.

Angles of Off-balancing

There are eight angles or directions in which an opponent can be off-balanced: forward, rear, right, left, forward right, forward left, rear right, and rear left.

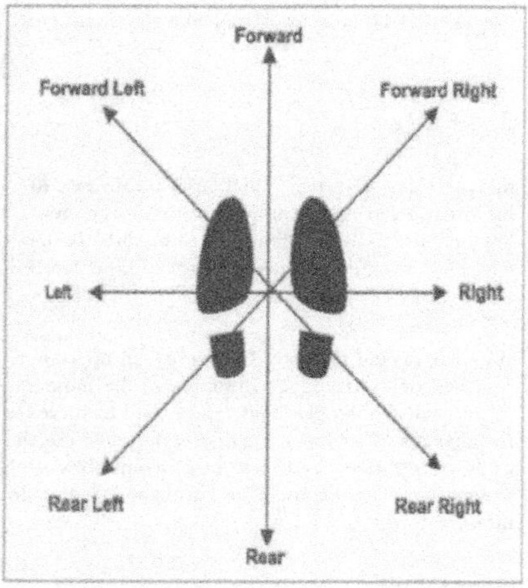

Note: The angles correspond to the Marine's perspective, not the opponent's. Forward, rear, right, and left are straight angles. Forward right, forward left, rear right, and rear left are considered quadrants that are at a 45-degree angle in either direction to the front or rear.

Off-balancing Techniques

Marines off-balance an opponent by pushing, pulling, or bumping the opponent with their hands, arms, or bodies.

To pull, Marines grab an opponent with their hands and drive him forcefully to one of the rear quadrants or to the right or left.

To push, Marines grab the opponent with their hands and drive him forcefully into one of the front quadrants or to the right or left.

Marines execute bumping in the same manner as pushing, but use their shoulders, hips, and legs instead of their hands to off-balance the opponent.

8. Falls

Marines may lose their balance or be thrown to the ground during encounters with an opponent. Marines use falling techniques to absorb the impact of a fall and to quickly return to their feet following an opponent's attack.

Whether falling or being thrown by an opponent, Marines strive to reduce the force of the impact, to prevent serious personal injury, and to increase the chances of survival. Falling techniques use the body's large muscles (back, thighs, buttocks) to protect vital organs and bones from injury and immobilization.

Front Fall

Marines execute a front fall to break a fall on the front. To execute the front fall, Marines-

Bend the elbows and place the palms facing out in a position to spread and absorb the impact of the fall.

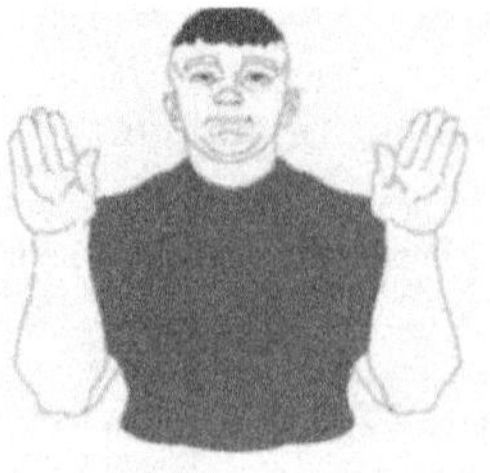

Fall forward, breaking the fall with the forearms and palms. The forearms and hands, down to the fingertips, should strike the ground simultaneousiy-

Offer resistance with the forearms and hands to keep the head raised off the ground.

Side Fall

Marines execute a side fall to break a fall on the side. To execute the side fall, Marines-

Bring the right arm across the body so the hand is next to the left shoulder with the palm facing inboard.

Fall to the side, breaking the fall with the right arm by slapping the ground and making contact from the shoulder or forearm down to the hand. At the same time, tuck the chin and keep the head raised off the ground. The chin should be tucked to the chest at all times to prevent whiplash.

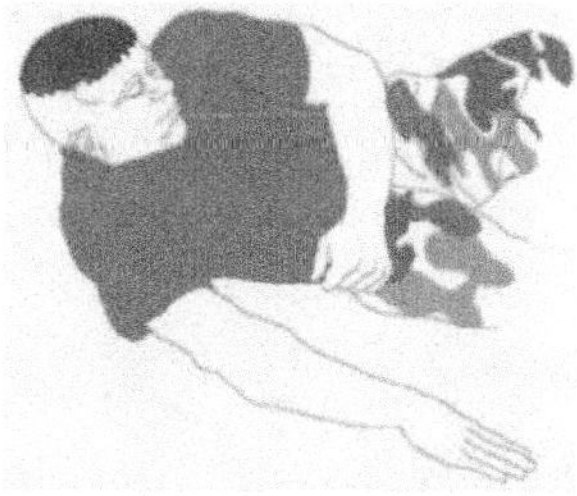

Stretch the right leg out to make contact with the ground and to distribute and absorb the impact. Bend the left leg, allowing the foot to make contact with the ground.

Back Fall

Marines execute a back fall to break the fall when being thrown or falling backward. To execute the backfall, Marines-

Cross the hands in front of the chest and tuck the chin.

Fall backward and slap the ground with the forearms and hands to absorb the impact of the fall and keep the head off the ground.

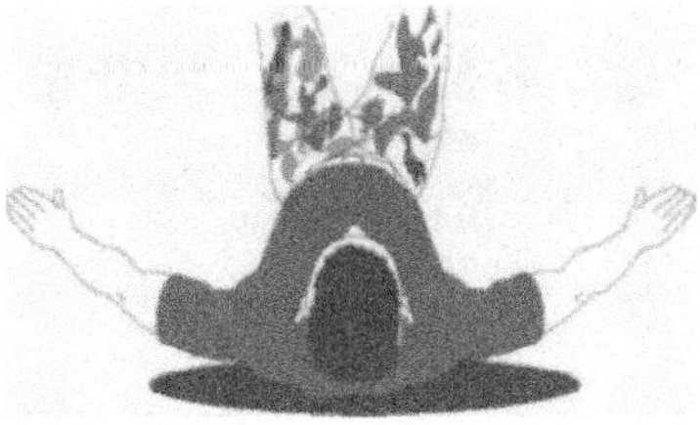

Forward Shoulder Roll

Marines use the forward shoulder roll to break a fall from an opponent's attack and to use the momentum of the fall to get back on their feet quickly. Ideally, Marines execute the forward shoulder roll to a standing position so they can continue fighting. To execute the forward shoulder roll, Marines-

Contact the ground with the back of the right forearm and upper arm. Tuck the chin into the chest.

Roll onto the right shoulder, rolling diagonally across the back to land on the left hip.

Slap the ground with the left arm, absorbing the impact from the shoulder to the hand, palm down.

Keep the left leg straight to absorb as much of the impact as possible. The right leg is bent and the foot hits flat on the ground.

Bend the left leg upon impact to push off with the left knee and leg to a squatting and then a standing position. Forward momentum should carry the Marine to a standing position.

CHAPTER 2

LETHAL AND NONLETHAL WEAPONS TECHNIQUES

This chapter describes all techniques for a right handed person. However, all techniques can be executed from either side.

In drawings, the Marine is depicted in woodland camouflage utilities; the opponent is depicted without camouflage. In photographs, the Marine is depicted in woodland camouflage utilities; the opponent is depicted in desert camouflage utilities.

1. Bayonet Techniques

WARNING

During training, Marines must have bayonets sheathed. Marines use bayonet dummies to practice bayonet techniques. When practicing offensive and defensive bayonet techniques student-on-student, Marines use pugil sticks.

All Marines armed with a rifle carry a bayonet. The bayonet is an effective weapon if Marines are properly trained in offensive and defensive bayonet techniques. An offensive attack, such as a thrust, is a devastating attack that can quickly end a fight. Defensive techniques, such as the block and parry, can deter the opponent's attack and allow Marines to regain the initiative. Through proper training, Marines develop the courage and confidence required to effectively use a bayonet to protect themselves and destroy the enemy. In situations where friendly and enemy troops are closely mingled and rifle fire and grenades are impractical, the bayonet becomes the weapon of choice.

Holding the Rifle

To execute bayonet techniques, Marines hold the rifle in a modified basic warrior stance. All movement begins and ends with the basic warrior stance. To hold the rifle, Marines—

Use an overhanded grasp to grab the small of the rifle's stock. Use an underhanded grasp to grab the handguards of the rifle.

Lock the buttstock of the rifle against the hip with the right forearm.

Orient the blade end of the rifle toward the opponent.

Offensive Bayonet Techniques

Straight Thrust. Marines use the straight thrust to disable or kill an opponent. It is the most deadly offensive technique because it causes the most trauma to an opponent. Target areas are the opponent' s throat, groin, or face. The opponent's chest and stomach are also excellent target areas if not protected by body armor or combat equipment. To execute the straight thrust, Marines—

Lift the left leg and lunge forward off the ball of the right foot while thrusting the blade end of the weapon forward, directly toward the opponent.

Retract the weapon and return to the basic warrior stance.

Slash. Marines use the slash to kill an opponent or to create an opening in his defense. The target area is the opponent's neck. To execute the slash, Marines—

Extend the left hand back toward the left shoulder.

Thrust the left hand forward and swing it to the right, bring the right hand back toward the hip, and turn the cutting edge of the blade toward the opponent's neck. The movement is a slashing motion so the blade cuts across the opponent's neck.

Horizontal Buttstroke. Marines use the horizontal buttstroke to weaken an opponent's defenses, to cause serious injury, or to set him up for a killing blow. Target areas are the opponent's head, neck, and legs. To execute the horizontal buttstroke, Marines—

Step forward with the right foot and drive the right hand forward. Rotate the hips and shoulders into the strike. Move the left hand back toward the left shoulder.

Strike the opponent with the butt of the weapon.

Vertical Buttstroke. Marines use the vertical buttstroke to weaken an opponent's defenses, to cause serious injury, or to set him up for a killing blow. Target areas are the opponent's groin and face. To execute the vertical buttstroke, Marines—

Step forward with the right foot and drive the right hand straight up.

Pull the left hand back over the left shoulder.

Strike the opponent with the butt of the weapon.

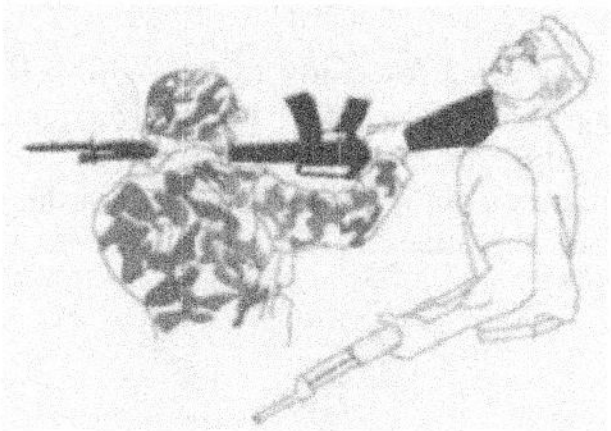

Smash. Marines use the smash as a follow-up technique to the horizontal or vertical buttstroke, particularly if they missed the target. The target area is the opponent's head. To execute the smash following a buttstroke, Marines-
Step forward with the right foot and place the blade end of the weapon over the left shoulder and elevate the right elbow above the shoulders.

Drive the arms straight forward, striking the opponent with the butt of the weapon.

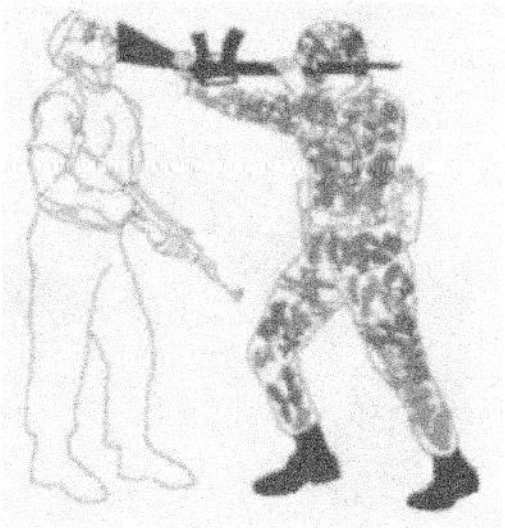

Defensive Bayonet Techniques

Parry. Marines use a parry as a defensive technique to redirect or deflect an attack. A parry is a slight redirection of a linear attack by an opponent; e.g., a straight thrust or a smash. To execute the parry, Marines-
Use the bayonet end of the rifle to redirect the barrel or bayonet of the opponent's weapon.

Lock the weapon against the hip with the right forearm.

Rotate to the right or left, moving the bayonet end of the rifle to parry the opponent's attack. Rotation is generated from the hips.

Redirect or guide the opponent's weapon away from the body by exerting pressure against the opponent' s weapon.

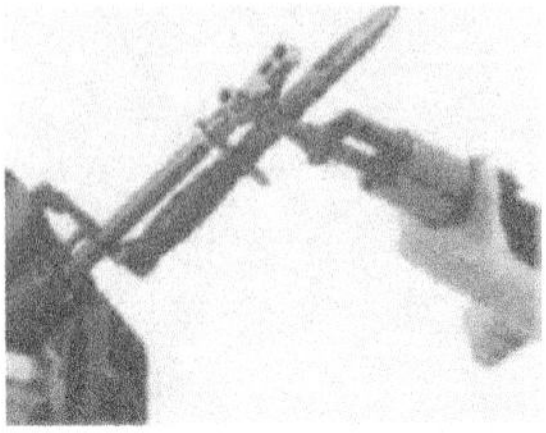

High Block. Marines execute a high block against a vertical attack coming from high to low. To execute the high block, Marines—

Thrust the arms up forcefully at approximately a 45-degree angle from the body. The weapon should be over the top of the head and parallel to the ground. The elbows are bent, but there is enough muscular tension in the arms to absorb the impact and deter the attack.

Low Block. Marines execute the low block

against a vertical attack coming from low to high. To execute the low block, Marines—

Thrust the arms down forcefully at approximately a 45-degree angle from the body. The weapon

should be at or below the waist and parallel to the ground. The elbows are bent, but there is enough muscular tension in the arms to absorb the impact and deter the attack.

Left and Right Block. Marines execute a left or right block against a horizontal buttstroke or a slash. To execute the left or right block, Marines-

Thrust the arms forcefully to the right or left, holding the rifle vertically in the direction of the attack. The elbows are bent, but there is enough muscular tension in the arms to absorb the impact and deter the attack.

Counter Action Following the Block. After deflecting an opponent's attack with a block, Marines counter with a slash or a horizontal buttstroke to regain the initiative. However, the objective in any bayonet fight is to thrust forward with the blade end of the weapon to immediately end the fight.

Group Strategy

On occasion, Marines may engage an opponent as a member of a group or numerous opponents by one's self or as a member of a group. By combining bayonet fighting movements and simple strategies, Marines can effectively overcome their opponent or opponents.

Offensive Strategy: Two Against One. If two bayonet fighters engage one opponent, the fighters advance together.

Fighter 1 engages the opponent while fighter 2 swiftly and aggressively attacks the opponent's exposed flank and destroys the opponent.

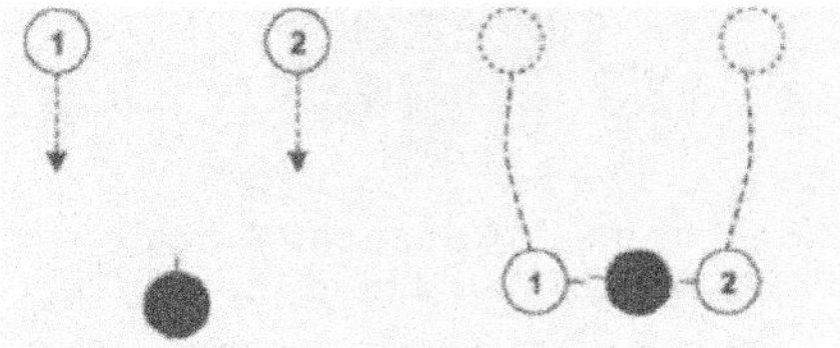

Offensive Strategy: Three Against Two. If three bayonet fighters engage two opponents, the fighters advance together keeping their opponents to the inside.

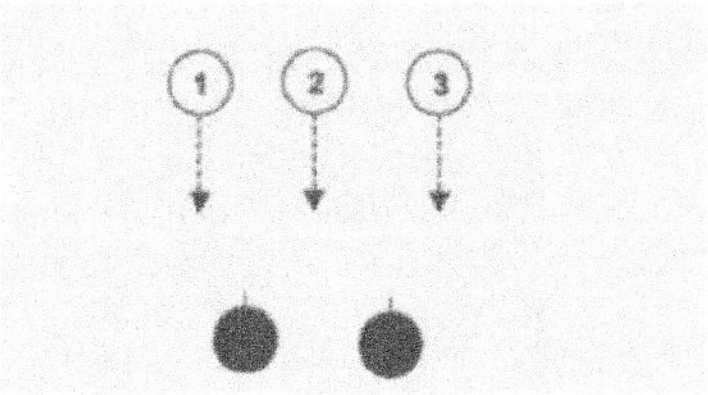

Fighters 1 and 3 engage opponents. Fighter 2 attacks the opponent's exposed flank engaged by fighter 1 and destroys the opponent.

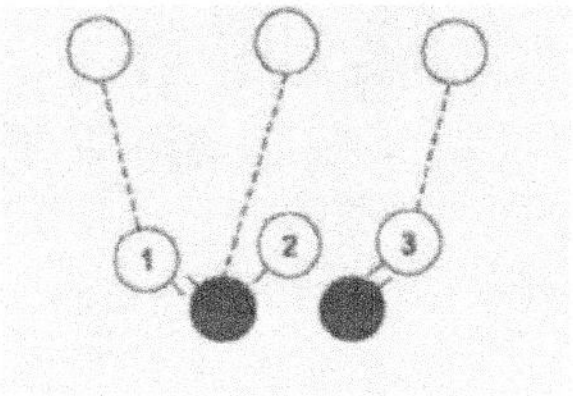

Fighters 1 and 2 turn and attack the exposed flank of the opponent engaged by fighter 3 and destroy the opponent.

Defensive Strategy: One Against Two. If a fighter is attacked by two opponents, the fighter immediately positions himself at the flank of the nearest opponent and keeps that opponent between himself and the other opponent.

Using the first opponent's body as a shield against the second opponent, the fighter destroys the first opponent quickly before the second opponent moves to assist.

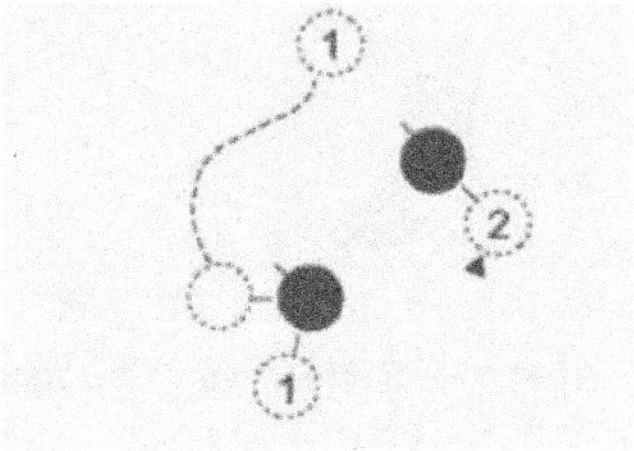

Then, the fighter engages and destroys the second opponent.

Defensive Strategy: Two Against Three. If two fighters are attacked by three opponents, the fighters immediately move to the opponent's flanks.

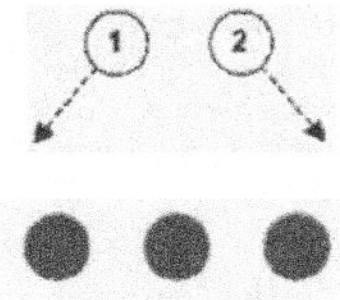

Fighters 1 and 2 quickly attack and destroy their opponents before the third opponent closes in.

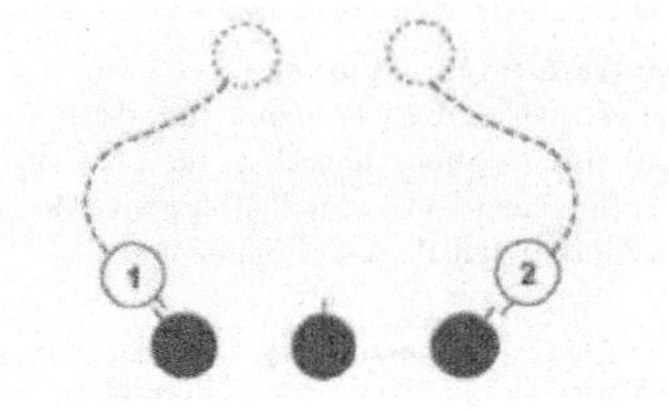

Fighter 1 engages the third opponent while fighter 2 attacks the opponent's exposed flank and destroys the opponent.

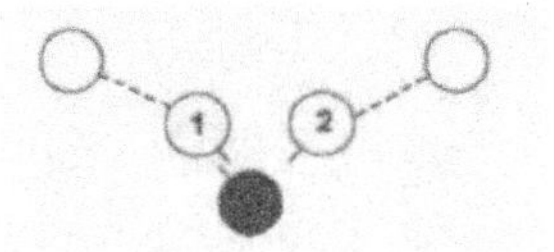

2. Nonlethal Rifle and Shotgun Retention Techniques

Most Marines are armed with the M16A2 service rifle. Marines are taught to keep their weapon with them at all times. Marines must be constantly alert to their surroundings and the people moving in and around their environment. Marines may be confronted by an individual who tries to take theii weapons. If this happens, Marines should not struggle with the individual. To retain positive control of their weapons, Marines must understand and apply weapons retention techniques, otherwise known as armed manipulation. The fol lowing techniques can be used with either the rifle or the shotgun.

Blocking Technique

To execute a blocking technique, Marines-

Stand in a defensive posture.

Use the weapon to block the opponent by thrusting it out firmly, with the elbows bent. Do not try to hit the opponent with the rifle. The rifle is used as a barrier.

Technique if Opponent Grabs Weapon Underhanded

Usually an opponent will try to grab the weapon or block it as an instinctive action. If the opponent uses an underhand grab to seize the handguards of the weapon, Marines—

Trap the opponent's closest finger(s) above the knuckle with the thumb so he cannot release his grip-

Apply bone pressure on the opponent's finger to initiate pain compliance.

Rotate the barrel of the weapon up or down quickly while maintaining pressure on the opponent's hand. At the same time, quickly pivot to off-balance the opponent.

Technique if Opponent Grabs Weapon Overhanded

If the opponent uses an overhand grab to seize the weapon overhanded, Marines—

Trap the opponent's finger to hold his hand in place.

Rotate the barrel to place it across the opponent's forearm and apply downward pressure. This action is similar to an armbar.

Technique if the Opponent Grabs the Muzzle

If the opponent grabs the muzzle of the weapon, Marines—

Rotate the muzzle quickly in a circle motion, Slash downward with the muzzle to release the opponent's grip.

Butt Strikes

Strikes with the butt of the weapon control or ward off an attacker. During any of the retention techniques, Marines use the heel or cutting edge of the weapon to deliver butt strikes to the inside or the outside of the opponent's thigh. The inside butt strike targets the femoral nerve as illustrated in the photo to the left.

The outside butt strike targets the peroneal nerve as illustrated in the photo to the left. Strikes can be made to the outside or inside of the thighs. If a strike to one side of the thigh misses, Marines follow back through with the butt of the weapon on the other side of the thigh.

Off-balancing Techniques

Marines apply off-balancing techniques to throw an opponent to the ground and retain possession of the weapon.

If the opponent grabs the weapon and pushes, Marines should not push on the weapon. They should—

Move with the momentum and movement of the

opponent by pivoting in the direction of the movement by stepping back.

Throw the opponent to the ground with a quick jerking movement by lowering the mu/./le and swinging the butt of the weapon.

If the opponent grabs the weapon and pulls, Marines—

Step on the opponent's foot and push forward to off-balance him and drive him to the ground.

Sweep the opponent's feet out from under him by hooking his leg with the leg and kicking backward.

3. Nonlethal Handgun Retention Techniques

Many Marines are armed with the M9 service pistol. Marines must keep their weapons in their possession at all times. Marines must be constantly alert to their surroundings and the people moving in and around the environment. Marines may be confronted by an individual who tries to take their weapons. To retain positive control of the weapon, Marines must understand and apply handgun retention techniques.

Blocking Technique

Marines perform the following blocking techniques if an opponent attempts to grab their pistol in the holster. Marines—

Place the body between the weapon and the opponent by immediately pivoting so the weapon is away from the opponent.

Step back and away from the opponent while placing the hand on the pistol grip.

Extend the left hand and block, deflect, or strike the opponent's arm.

Armbar Technique

Marines use the armbar technique when an opponent uses his right hand to grab their pistol in the holster. To execute the armbar technique, Marines—

Trap the opponent's right hand by grasping the opponent's wrist or hand with the right hand and applying pressure against the body.

Step back with the right foot and pivot sharply to the right to be next to the opponent. Always pivot in a direction that keeps the weapon away from the opponent.

Straighten the opponent's arm to apply an armbar. The arm should be straight across the torso.

Continue pivoting to the right while pulling back on the opponent's shoulder. This action may break the opponent's arm.

Place the left hand across the opponent's face and apply pressure back and down to take the oppo-

nent to the ground. Pressure applied beneath the nose or on the trachea is equally effective.

Wristlock Technique

Marines use the wristlock technique when an opponent grabs their pistol while it is in the holster with his right hand. To execute the w ristlock technique, Marines—

Grasp the opponent's wrist or hand with the right hand and apply pressure against the body.

Step back and away from the opponent and pivot to the right so the weapon is away from the opponent. Always pivot in a direction that keeps the weapon away from the opponent.

Reach over the opponent's arm with the left hand and grab his hand, applying pressure against his arm with the left forearm. Execute a wristlock.

Incorporate the second hand into the wristlock and, stepping back with the left foot, pivot to the left.

Execute a two-handed wristlock by exerting downward pressure with the thumb and rotating his hand to the left.

Continue to pivot to off-balance the opponent and drive him to the ground.

Note: If the opponent grabs the pistol with his left hand, Marines execute the wristlock with one hand and step in toward the opponent, rather than

away from the opponent.

Softening Techniques

Handgun retention techniques use softening techniques applied to pressure points. Bone pressure

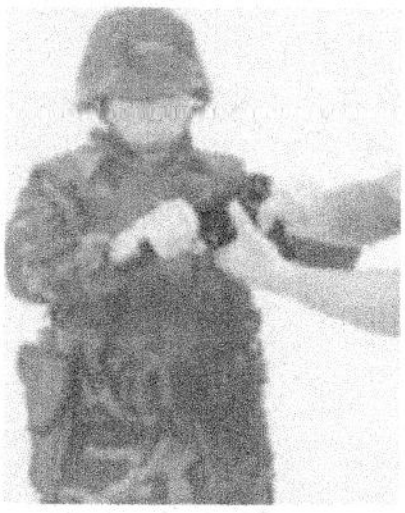

and strikes with the hands (i.e., hammer fist), knees, and feet are also effective softening techniques.

Pressure Points. Marines use pressure point techniques to get the opponent to loosen his grip. Marines use their finger tips to apply pressure to the webbing between the index finger and thumb, the jugular notch, and the brachial plexus tie in. The following figure illustrates pressure applied to the brachial plexus tie in.

Bone Pressure. Bone pressure is the application of pressure on a bone against a hard object to initiate pain compliance. To apply bone pressure, Marines use their hand to trap the opponent's hand on the weapon. Marines apply a slow, steady pressure to the opponent's hand and fingers until his grip is softened or he releases his hold.

Strikes. If it is difficult to apply a retention technique, Marines employ strikes or kicks to force the opponent to loosen his grip. Strikes to the

eyes, the arms (radial nerve), or shoulder (brachial plexus tic in) soften the opponent's grip on the weapon.

Kicks and knee strikes to the peroneal nerve, the femoral nerve, or the groin arc effective because the opponent is typically unprepared to counter the strike.

Stomping on the top of the opponent's foot may distract him or loosen his grip on the weapon.

4. Firearm Disarmament Techniques

Marines use firearm disarmament techniques during a close-range confrontation if they are un-

armed and the opponent has a firearm (pistol). These techniques are equally effective if Marines are armed but do not have time to withdraw and present the weapon. The goal of firearm disarmament techniques is to gain control of the situation

so Marines gain the tactical advantage. The goal is not necessarily to get control of the opponent's weapon.

Pistol to the Front

This technique is used when Marines are unarmed and the opponent has a pistol pointing at their front (e.g., head, chest). The technique is the same if the opponent sticks the pistol under the Marine's chin. To execute the counter when an opponent is pointing a pistol toward the front of a Marine, Marines—

Place the hands close to the weapon, about chest high, palms out.

Use the left hand to grab the opponent's forearm and push the opponent's hand with the pistol away to clear the body from in front of the weapon. At the same time, rotate the right shoulder back to clear the body from the weapon.

Maintain control of the opponent's arm.

Grasp the weapon with the right hand by placing the thumb underneath the pistol and the fingers over top of the pistol.

Keep the right hand wrapped tightly around the muzzle and quickly rotate the pistol in the opponent's hand so the muzzle is facing the opponent.

Grasp and pull the opponent's wrist or forearm away from the body while rotating the weapon.

Rotate the weapon toward the opponent while pulling it up and back and out of the opponent's grasp.

Pistol to the Rear

This technique is performed when Marines are unarmed and the opponent has a pistol pointing to the back of the Marine's head. To execute the counter when the opponent is pointing a pistol to the rear, Marines—

Place the hands close together about chest high, palms out.

Step back with the left foot, pivoting on the right foot so the side is against the opponent's front. This action clears the body from the weapon's line of fire. Keep the left hand up.

Pivot on the left foot to face the opponent and, at the same time, raise the left elbow and reach over the top of the opponent's arm with the left arm.

Wrap the left arm tightly around the opponent's arm above his elbow to control it.

Push on the opponent's shoulder with the right hand while pulling up with the left arm to achieve an armbar. This action releases the opponent's grip on the weapon. If necessary, the opponent can be taken to the ground with a leg sweep.

Note: To execute this technique, the weapon must be close to or touching the rear of the Marine. If the weapon is too far away from the body, this technique would be difficult to execute or it would be ineffective.

CHAPTER 3

HAND-HELD WEAPONS

This chapter describes all techniques for a right-handed person. However, all techniques can be executed from either side.

In drawings, the Marine is depicted in woodland camouflage utilities; the opponent is depicted without camouflage. In photographs, the Marine is depicted in woodland camouflage utilities; the opponent is depicted in desert camouflage utilities.

Marines must know how to defend against attacks when an opponent is either unarmed or armed with a held-held weapon. This chapter addresses the combative use of knives, specific weapons of opportunity, and sticks. However, virtually anything can be used as a hand-held weapon.

1. Fundamentals of Knife Fighting

Marines must be trained in knife fighting techniques. Marines experienced in offensive knife techniques can cause enough damage and massive trauma to stop an opponent. When engaged against each other, experienced knife fighters employ various maneuvers and techniques that are specific to knife fighting. Seldom, if ever, will Marines engage an opponent in a classical knife fight.

Note: When armed with a rifle, Marines are issued a bayonet. When armed with a pistol, Marine are issued a combat knife.

Angles of Attack

There are six angles from which an attack with a knife can be launched:

- Vertical strike coming straight down on an opponent.
- Forward diagonal strike coming in at a 45-degree angle to the opponent.
- Reverse diagonal strike coming in at a 45-degree angle to the opponent.
- Forward horizontal strike coming in parallel to the ground.
- Reverse horizontal strike coming in parallel to the ground.
- Forward thrust coming in a straight line to the opponent.

Target Areas of the Body

During any confrontation, the parts of the opponent's body that are exposed or readily accessible will vary. The goal in a knife fight is to attack the body's soft, vital target areas that are readily accessible (e.g, the face, the sides and front of the neck, the lower abdomen [or groin]).

Neck. Carotid arteries, located on either side of the neck, are good target areas because they are not covered by body armor or natural protection.

Lower Abdomen (or Groin). The lower abdomen (or groin region) is a good target area because it is not covered by body armor.

Heart. The heart, if not covered by body armor, is an excellent target which, if struck, can prove fatal in a matter of seconds or minutes.

Secondary Targets. There are secondary target areas that will cause substantial bleeding if an artery is severed. These target areas are not immediately fatal, but can become fatal if left unattended. Attacks to—

> The legs can cause a great deal of trauma and prove fatal. For example, the femoral artery located on the inside of the thigh is a large artery which, if cut, will cause extensive blood loss.
> The brachial artery, located between the biceps and triceps on the inside of the arm, can cause extensive bleeding and damage.
> The arm's radial and ulnar nerves can cause extensive bleeding and damage.

Movement

Marines can move anywhere within a 360-degree circle around the opponent. This allows accessibility to different target areas of the opponent's body. Marines should avoid being directly in front of an opponent because the opponent can rely on his forward momentum to seize the tactical advantage. If Marines face an opponent, movement is made in a 45-degree angle to either side of the opponent. This angle avoids an opponent's strike and places Marines in the best position to attack an opponent.

Wearing the Combat Knife

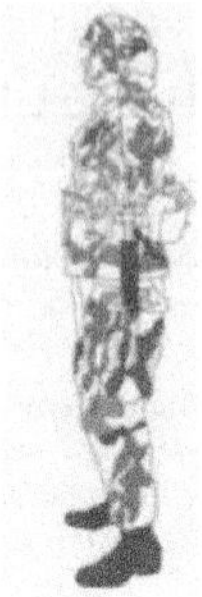

Marines must wear the combat knife where it is easily accessible and where it can best be retained. It is recommended the combat knife be worn on the weak side hip, blade down, sharp edge facing forward. Marines can place it behind the magazine pouch where it is easily accessible to them, but not easily grabbed by an opponent.

Grip

The grip on the knife should be natural. Marines grasp the knife's grip with the fingers wrapped around the grip naturally as it is pulled out of its sheath. This is commonly known as a hammer grip. The blade end of the knife is always facing the opponent.

Stance

Marines use the basic warrior stance as the foundation for knife techniques. The left hand is a vertical shield that protects either the ribs or the head and neck. The right elbow is bent with the blade pointing forward toward the opponent' s head. This position serves as an index point, where all techniques are initiated.

Principles of Knife Fighting

The following are key principles of knife fighting:

Execute movements with the knife blade within a box, shoulder-width across from the neck down to the waistline. The opponent has a greater chance of blocking an attack if the blade is brought in a wide, sweeping movement to the opponent.
Close with the opponent, coming straight to the target.
Move with the knife in straight lines.
Point the knife's blade tip forward and toward the opponent.
Apply full body weight and power in each of the knife techniques. Full body weight should be put into the attack in the direction of the blade's movement (slash or thrust).
Apply constant forward pressure with the body and blade to keep the opponent off-balanced.

2. Knife Fighting Techniques

Slashing Techniques

Marines use slashing techniques to close with an enemy. Slashing techniques distract the opponent or damage the opponent so Marines can close in. Typically, Marines target the opponent's limbs, but any portion of the body that is presented can become a target.

Vertical Slash Technique. The vertical slash follows a vertical line straight down through the target. To execute the vertical slash, Marines—

Thrust the right hand out and bring the weapon straight down on the opponent, continuing to drag the knife down through the opponent's body.

Maintain contact on the opponent's body with the blade of the knife.

Forward Slash Technique. The forward slash follows a straight line in a forehand stroke, across the target areas of either the neck (high diagonal slash) or abdominal region (low horizontal slash). To execute the forward slash, Marines-
Extend the right hand while simultaneously rotating the palm up until the knife blade makes contact with the opponent.

Snap or rotate the wrist through the slashing motion to maximize blade's contact with the opponent.

Drag the knife across the opponent's body, from right to left, in a forehand stroke. The movement ends with the forearm against the body and the knife at the left hip with its blade oriented toward the opponent.

Reverse Slash Technique. The reverse slash is a follow-up technique to a forward attack. It allows Marines both a secondary attack and the ability to resume the basic warrior stance. The reverse slash follows a straight line in a backhand stroke, across the target areas of either the neck (high diagonal slash) or abdominal region (low horizontal slash). To execute the reverse slash, Marines—

Extend the right hand while simultaneously rotating the palm down until the knife blade makes contact with the opponent.

Snap or rotate the wrist through the slashing motion to maximize the blade's contact with the opponent.

Drag the knife across the opponent's body, from left to right, in a backhand stroke. Maintain contact on the opponent's body with the blade of the knife.

Thrusting Techniques

The primary objective of knife fighting is to insert the blade into an opponent to cause extensive damage and trauma. This is done with a thrusting technique. Thrusting techniques are more effective than slashing techniques because of the damage they can inflict. However, Marines use slashing techniques to close with the enemy so that they are closer to the opponent, which allows them to use the thrusting technique.

Vertical Thrust. The thrusting motion follows a vertical line straight up through the target (low into the abdomen region or high into the neck). To execute the vertical thrust, Marines—

Thrust the right hand toward the target, inserting the knife blade straight into the opponent.

Pull the knife out of the opponent.

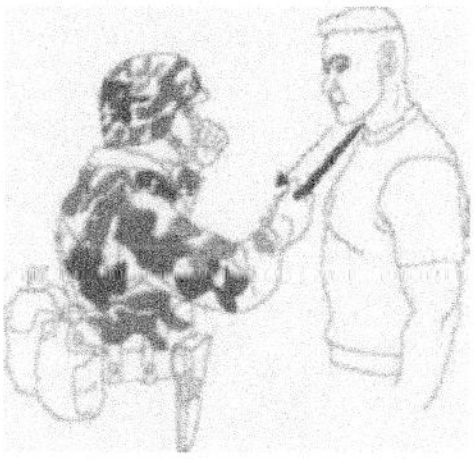

Forward Thrust. The forward thrust follows a straight line straight into the opponent's neck (high thrust) or abdominal region (low thrust). To execute the forward thrust, Marines—

Thrust the right hand, palm down, toward the target, inserting the knife blade straight into the opponent.

Rotate the palm up once the knife is inserted to twist the blade.

Drop the right elbow and bring the knife to the opposite side of the opponent's body from where it was inserted. At the same time, rotate the hips and shoulders downward to bring body weight to bear on the attack.

Reverse Thrust. The reverse thrust is a follow-up technique to a forward attack. It allows Marines both a secondary attack and the ability to resume the basic warrior stance. The reverse thrust follows a horizontal line straight into the opponent's neck (high thrust) or abdominal region (low thrust). To execute the reverse thrust, Marines-

Bend the right arm, crossing the arm to the left side of the body.

Thrust the right hand, palm up, toward the target, inserting the knife blade straight into the opponent.

Rotate the palm down to twist the blade once the knife is inserted.

Bring the knife to the opposite side of the opponent's body from where it was inserted.

3. Weapons of Opportunity

During an unarmed close combat situation, Marines use their bodies as weapons, but they should be ready and able to use anything around them as a weapon. For example, Marines could throw sand or liquid in an opponent's eyes to temporarily impair his vision or smash the opponent's head with a rock or helmet. Marines must use whatever means are available and do whatever it takes to take control of the situation and to win, or they face the possibility of losing their lives. Some weapons of opportunity are discussed in the following subparagraphs.

Entrenching Tool

An entrenching tool (E-tool) is commonly carried by Marines. It can be an excellent weapon, especially when sharpened. Marines can use the E-tool to block, slash, and thrust at an opponent.

Tent Pole and Pins

Marines can use tent poles and pins to block, strike, or thrust at an opponent.

Web Belt

Marines can stretch a web belt between their hands to block attacks by an opponent.

Battlefield Debris

Marines can use debris on the battlefield (e.g., sticks, glass, a sharp piece of metal) to cut, slash, or stab an opponent. They can also use other types of debris such as shovel or ax handles, boards, metal pipes, or broken rifles to strike an opponent or apply a choke.

Helmet

A helmet can be used to strike an opponent on an unprotected area like the head and face. Grasp the rim of the helmet and thrust the arms forward, striking the opponent with the top of the helmet.

4. Fundamentals of Combative Stick

On the battlefield, Marines must be ready and able to use anything as a weapon. They must learn and be able to use techniques that can be employed with most weapons of opportunity. Among these techniques are combative stick techniques. Combative stick techniques can be used with a stick, a club, a broken rifle, an E-tool, or even a web belt.

Angles of Attack

There are six angles from which an attack with a hand-held weapon can be launched:

Vertical strike coming straight down on an opponent.
Forward diagonal strike coming in at a 45-degree angle to the opponent.
Reverse diagonal strike coming in at a 45-degree angle to the opponent.
Forward horizontal strike coming in parallel to the ground.
Reverse horizontal strike coming in parallel to the ground.
Forward thrust coming in a straight line to the opponent.

Grip

Grasp the stick about 2 inches from its base.

Stance

The basic warrior stance serves as the foundation for combative stick techniques. The left hand becomes a vertical shield that protects the ribs or the head and neck. Depending on how heavy the weapon is, it should be held at a level approximately shoulder height.

Movement

Movement during combative stick techniques is the same as it is for other close combat techniques. Marines can move anywhere within a 360-dcgrcc circle around the opponent. This allows accessibility to different target areas of the opponent's body and gains a tactical advantage. Marines should avoid being directly in front of an opponent because the opponent can rely on his forward momentum to seize the tactical advantage. If Marines face an opponent, movement is made in a 45-dcgrcc angle to cither side of the opponent. This angle avoids an opponent's strike and places Marines in the best position to attack an opponent.

5. Combative Stick Techniques

Strikes

Strikes are intended to inflict as much damage on an opponent as possible. Striking techniques apply to a weapon of opportunity such as a stick, a tent pole, a club, a broken rifle, an E-tool, or a Pipe-

Vertical Strike. To execute the vertical strike, Marines—

Bend the right arm, extending the weapon over the back of the right shoulder.

Rotate the forearm straight down off the elbow to bring the weapon down on the opponent.

Rotate the hips and shoulders forcefully toward the opponent.

Follow through with the strike by allowing the weight of the weapon to go through the target area of the body.

Forward Strike. To execute a forward strike, Marines—

Step forward with the left foot in the direction of the strike.

Bend the right arm with the elbow extending out to the right and the weapon extended over the right shoulder.

Rotate the forearm to the right of the elbow to bring the weapon down onto the opponent. At the same time, forcefully rotate the hips and shoulders toward the opponent.

Follow through with the strike by allowing the weight of the weapon to go through the target area of the body.

Reverse Strike. The reverse strike is a follow-up technique to a forward strike. It allows Marines both a secondary attack and the ability to resume the basic warrior stance.To execute a reverse strike, Marines—

Step forward with the right foot in the direction of the strike.

Bend the right arm with the hand near the left shoulder. The weapon is extended over the left shoulder.

Rotate the forearm to the right of the elbow to bring the weapon down onto the opponent. At the same time, forcefully rotate the hips and shoulders toward the opponent.

Follow through with the strike by allowing the weight of the weapon to go through the target area of the body.

Forward Thrust. To execute the forward thrust, Marines—

Grasp the stick with the left hand, palm up, in a position where the stick can be controlled with two hands.

Lift the left leg and lunge forward off the ball of the right foot. At the same time, thrust the end of the weapon directly toward the opponent by thrusting both hands forward in a straight line.

6. Blocking Techniques

A block is meant to deter or deflect an attack by an opponent. A block sets up Marines for a follow-on attack against the opponent. Blocks are executed by deflecting, rather than hitting or following through like a strike.

Blocks Against Unarmed Attacks

To block against an unarmed attack, Marines—

Step forward at a 45-degree angle with either the right or left foot. This moves the body out of the line of attack.

Raise the left arm and block the strike with the meaty portion of the forearm.

Employ the stick by one of two techniques:

Strike the opponent with the stick.
Use the stick to block on two points of contact. When the stick is used to block, it serves as an incidental strike. This technique is only used if Marines have closed with the opponent and are inside his strike.

Blocks Against Armed Attacks

Block for a Vertical Strike. To execute the block for a vertical strike, Marines—

Step forward with the left foot at a 45-degree angle to the left. This moves the body out of the line of attack.

Block on two points of contact to disperse the impact of the attack:

Block the opponent's stick by positioning the stick so it is perpendicular to the opponent's stick. If the stick is not perpendicular to the opponent's stick, the stick can slide through and make contact on the Marine.
Block the opponent's wrist or forearm with the back of the left forearm.

Note: Use the stick to block the opponent's arm if closer to the opponent. It is the same movement as blocking with the arm, except the opponent's arm is blocked with both the stick and the arm.

Block for a Forward Strike. To execute the block for a forward strike, Marines-
Step forward with the left foot at a 45-degree angle to the left. This moves the body out of the line of attack and inside the opponent's strike.

Block on two points of contact to disperse the impact of the attack:

- Block the opponent's wrist or forearm with the meaty portion of the back of the left forearm.
- Strike the opponent's attacking biceps with the stick.

Block for a Reverse Strike. To execute the block for a reverse strike, Marines—

Step forward at a 45-degree angle with the right foot to the right. This moves the body out of the line of attack and inside the opponent's strike.

Block on two points of contact to disperse the impact of the attack:

Block the opponent's stick by positioning the stick so it is perpendicular to the opponent's stick.

Block the opponent's forearm with the meaty portion of the left forearm.

Note: If closer to the opponent, block his triceps with the back of the left forearm and strike his forearm with the stick.

7. Unarmed Against Hand-Held Weapons

If Marines are engaged against an opponent with a knife, a stick, or some other weapon of opportunity, they must establish and maintain an offensive mindset, not a defensive mindset. Their survival depends on it. They cannot afford to think about getting cut or hurt.

Angles of Attack

Before Marines can learn to block or counter an attack with a hand-held weapon (e.g., knife, stick), they must know from what angle the opponent is attacking. There are six angles from which an opponent will typically attack with a hand-held weapon:

Vertical strike coming in straight down on the Marine.
Forward diagonal strike coming in at a 45-degree angle to the Marine.
Reverse diagonal strike coming in at a 45-degree angle to the Marine.
Forward horizontal strike coming in parallel to the ground.
Reverse horizontal strike coming in parallel to the ground.
Forward thrust coming in a straight line to the Marine.

Blocks

If the opponent has a hand-held weapon (e.g., knife or stick), Marines parry the opponent's hand or arm to block the attack.

Basic Block. To execute the basic block technique, Marines-

Step forward at a 45-degree angle to move out of the line of the attack. Always step in the direction of the strike.

Thrust the forearms forward, hands up, against the opponent's attacking arm. Contact is made on the opponent's arm with the backs of the forearms.

Block for a Vertical Strike. To execute the block against a vertical strike, Marines—

Step forward with the left foot at a 45-degree angle to the left to move out of the line of attack.

Thrust the forearms forward, hands up, against the outside of the opponent's attacking arm.

Block for a Forward Strike. To execute the block against a forward diagonal or forward horizontal strike, Marines-

Step forward with the left fool inside the opponent's attacking arm.

Block the attack with both arms bent so the forearms make contact with the opponent's biceps and forearm.

Block for a Reverse Strike. To execute the block against a reverse diagonal or reverse horizontal strike, Marines—

Step forward with the right foot to the outside of the opponent's attacking arm.

Block the attack with both arms bent so the forearms make contact with the opponent's triceps and forearm.

Block for a Forward Thrust. To execute the block against a forward thrust, Marines—

Bend at the waist, move the hips backward, and jump backward with both feet to move away from the attack. This action is known as "hollowing out."

Hollow out and block the attack with the arms bent and hands together on top of the attacking arm.

Overlap the hands slightly so one thumb is on top of the other hand's index finger. The other thumb should be under the other hand's index finger.

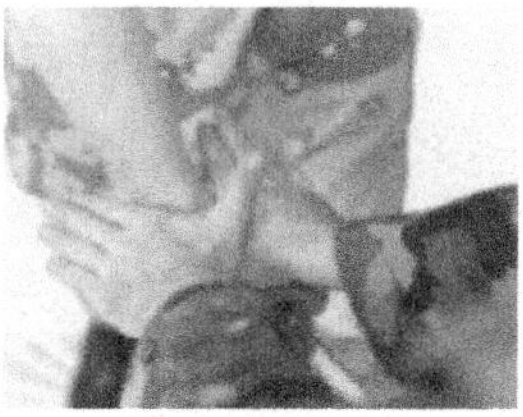

8. Counters to Hand-Held Weapon Attacks

Principles of Counters

A counter is used to control the situation to regain the tactical advantage and end the fight. Regardless of the type of weapon or angle of attack, the following principles apply to countering the attack with a hand-held weapon:

Move out of the line of attack. Movement is executed in a 45-degree angle forward to the left or right.

- Block the attack.

Note: The first two actions are taken simultaneously.

- Take control of the weapon by controlling the hand or arm that is holding the weapon. Never attempt to grab the opponent's weapon.
- Execute the appropriate follow-up to end the fight; e.g., strikes, joint manipulations, throws, or takedowns (see chap. 8). Marines should continue to attack the opponent until the fight ends.

Counter Techniques

There are two techniques that can be used to counter any armed attack: forward armbar counter and reverse armbar counter. These techniques can be used to counter a vertical attack, a forward diagonal strike, or a forward horizontal strike. With minor variations, the same techniques are used to counter reverse strikes. A third technique, the bent armbar counter, is used to counter a vertical attack.

Forward Armbar Counter. To execute the forward armbar counter to an attack coming from a forward strike, Marines-

Step forward with the left foot inside the opponent' s attacking arm.

Block the attack with both arms bent so the forearms make contact with the opponent's biceps and forearm.

Slide the left arm over the opponent's forearm and wrap the arm tightly around his arm, trapping the opponent's attacking arm between the biceps and torso.

Place the right hand on the opponent's shoulder or upper arm to further control his arm and to effect an armbar.

Execute an armbar and continue to exert steady pressure against the arm.

Reverse Armbar Counter. To execute the reverse armbar counter to an attack coming from a forward strike, Marines-

Step forward with the left foot inside the opponent' s attacking arm.

Block the attack with both arms bent so the forearms make contact with the opponent's biceps and forearm.

Control the opponent's arm with the left hand and pivot to the right so the back is against the opponent's side. Immediately slide the right arm over the opponent's biceps and wrap the arm tightly around his arm, trapping the opponent's attacking arm between the biceps and torso.

Grasp the opponent's wrist with the left hand and twist his thumb away from the body.

Take the opponent to the ground with an armbar takedown if he does not drop the weapon from the wristlock.

Bent Armbar Counter. This counter is particularly effective against a vertical attack. To execute the bent armbar counter, Marines—

Step forward with the left foot inside the opponent's attacking arm.

Block the attack with both arms bent so the forearms make contact with the opponent's biceps and forearm.

Grasp the opponent's forearm with the left hand. At the same time, slide the right arm underneath the opponent's triceps and grasp the opponent's forearm or wrist with the right hand.

Apply downward pressure with the hands against the opponent's forearm to off-balance the opponent.

CHAPTER 4

STRIKES

This chapter describes all techniques for a right-lmnded person. However, all techniques can be executed from either side.

In drawings, the Marine is depicted in woodland camouflage utilities; the opponent is depicted without camouflage. In photographs, the Marine is depicted in woodland camouflage utilities; the opponent is depicted in desert camouflage utilities.

Strikes are unarmed individual hitting techniques. Strikes use the hands, elbows, knees, feet, and, in some instances, other parts of the body as personal weapons. Marines must know how to execute strikes effectively. They must also know how to counter strikes from an opponent.

1. Principles of Punches

Muscle Relaxation

Muscle relaxation is crucial when executing punches. The natural tendency in a fight is to tense up, which results in rapid fatigue and decreased power generation. Marines who remain relaxed during a close combat situation generate greater speed, which results in greater generation of power. Relaxing the forearms generates speed and improves reaction time. At the point of impact, Marines clench the fist to cause damage to the opponent and avoid injury to the wrist and hand.

Weight Transfer

Weight transfer is necessary to generate power in a punch. Marines accomplish this by-

Rotating their hips and shoulders into the attack.

Moving their body mass straight forward or backward in a straight line.

Dropping their body weight into an opponent. Body mass can be transferred into an attack from high to low or from low to high.

Rapid Retraction

When Marines deliver a punch, rapid retraction of the fist is important. Once the hand has made contact with the target, Marines quickly return to the basic warrior stance. Rapid retraction-

Returns the hand and arm to the protection afforded by the basic warrior stance.

Prevents the opponent from being able to grab the hand or arm.

Permits the hand and arm to be "chambered" or "re-cocked" in preparation for delivering a subsequent punch.

Telegraphing

Telegraphing a strike occurs when body movements inform the opponent of the intention to launch an attack. Staying relaxed helps to reduce telegraphing.

Often, an untrained fighter telegraphs his intention to attack by drawing his hand back in view of his opponent, changing facial expression, tensing neck muscles, or twitching. These movements, however small, immediately indicate an attack is about to be delivered. If the opponent is a trained fighter, he may be able to evade or counter the attack. If the opponent is an untrained fighter, he may still be able to minimize the effect of an attack.

2. Punches

Punches may be thrown during any hand-to-hand confrontation. Most people resort to punching because it is a natural reaction to a threat. The purpose of a punch is to stun the opponent or to set him up for a follow-up finishing technique. However, punches should only be executed to the soft tissue areas of an opponent. A correctly delivered punch maximizes the damage to an opponent while minimizing the risk of injury to Marines.

Basic Fist

Punches are executed using the basic fist. To make the basic fist, the fingers are curled naturally into the palm of the hand and the thumb is placed across the index and middle fingers. Do not clench the fist until movement has begun. This reduces muscular tension in the forearm and increases speed and reaction time. Just before impact, Marines exert muscular tension on the hand and forearm to maximize damage to the opponent and reduce their chances of injury. Contact should be made with the knuckles of the index and middle fingers.

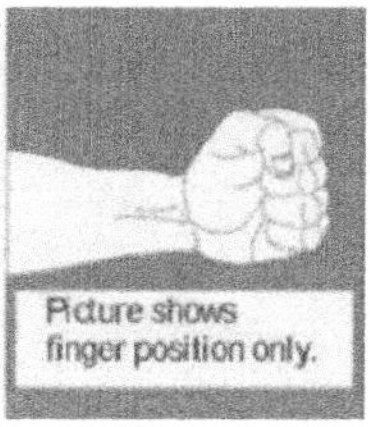

Picture shows finger position only.

When striking with the basic fist, Marines must keep the hand straight, or in line, with the wrist to avoid injury to the wrist.

Lead Hand Punch

The lead hand punch is a snapping straight punch executed by the forward or lead hand. It is a fast punch designed to keep the opponent away and to set up an attack. A lead hand punch conceals movement and allows Marines to get close to the opponent. Lead hand punches should strike soft tissue areas, if possible. To execute the lead hand punch, Marines-

Snap the lead hand out to nearly full extension, while rotating the palm to the ground.

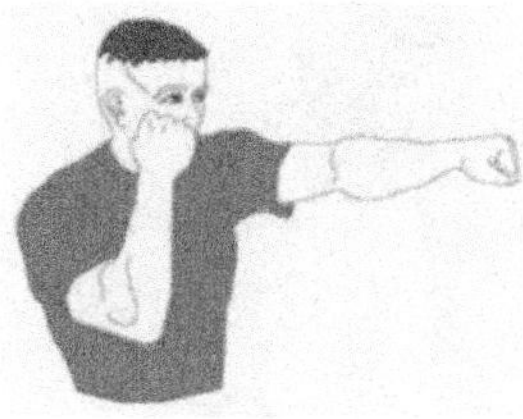

Contact the opponent with the first two knuckles of the fist.

Retract the hand immediately, resuming the basic warrior stance.

Rear Hand Punch

The rear hand punch is a snapping punch executed by the rear (right) hand. It is a power punch designed to inflict maximum damage on the opponent. Its power comes from pushing off the rear leg and rotating the hips and shoulders. To execute the rear hand punch, Marines—

Rotate the hips and shoulders forcefully toward the opponent and thrust the rear hand straight out, rotating the palm down, to nearly full extension.

Shift body weight to the lead foot while pushing off on the ball of the rear foot.

Contact the opponent with the first two knuckles of the fist.

Retract the hand immediately.

Uppercut

The uppercut is a powerful punch originating below the opponent's line of vision. It is executed in an upward motion traveling up the centerline of the opponent's body. It is delivered in close and usually follows a preparatory strike that leaves the target area unprotected. When delivered to the chin or jaw, the uppercut can render an opponent unconscious, cause extensive damage to the neck, or sever the tongue. To execute the uppercut, Marines—

Bend the arms, rotating the palm inboard. The distance the arms bend depends on how close the opponent is.

Rotate the hips and shoulders forcefully toward the opponent, thrusting the fist straight up toward the opponent's chin or jaw.

Contact the opponent with the first two knuckles of the fist.

Retract the hand immediately.

Hook

The hook is a powerful punch that is executed close in and is usually preceded by a preparatory strike. To execute the hook, Marines—

Thrust the right arm in a hooking motion toward the opponent, keeping the elbow bent while forcefully rotating the right shoulder and hip toward the opponent.

Contact the opponent with the first two knuckles of the fist. Continue rotating the shoulder and hip, following through with the fist to the target.

Retract the hand immediately.

3. Strikes with the Upper Body

Strikes stun the opponent or set him up for a follow-up finishing technique. The hands, forearms, and elbows are individual weapons of the arms that can be used to execute strikes including the hammer fist, chin jab, knife hand, eye gouge, and elbow strikes. These strikes provide a variety of techniques that can be used in any type of close combat situation.

Principles of Execution

There are several principles of execution that ensure a strike's effectiveness.

Generating Power. Weight transfer is necessary to generate maximum power in a punch. Marines accomplish this by-

Rotating the hips and shoulders into the attack.
Moving the body mass straight forward or backward in a straight line.
Dropping body weight into an opponent. Body mass can be transferred into an attack from high to low or from low to high.

Muscular Tension. The arms are relaxed until the moment of impact. At the point of impact, Marines apply muscular tension in the hand and forearm to maximize damage to the opponent and to avoid injury to the hand. The arms are relaxed until the moment of impact.

Hit and Stick and Follow-Through. A strike should be delivered so that the weapon (e.g., hand, elbow) hits and remains on the impact site (target) and follows through the target. This technique inflicts maximum damage on the opponent.

Strikes with the arms are executed with "heavy hands"; i.e., the strike is executed by driving through with the strike to allow the weight of the hand to go through the target area of the body. Contact with the opponent should be made with the arm slightly bent, and the arm extends as it moves through the target. This technique allows Marines to deliver strikes effectively without executing full force.

Movement. Movement puts Marines in the proper position for launching an attack against an opponent as well as providing protection.

Movement is initiated from the basic warrior stance and ends by resuming the basic warrior stance. Strikes can be performed with either the left or right arm depending upon—

The angle of attack.
The position of the opponent.
The opponent's available, vulnerable target areas.

Target Areas of the Body

For each strike, there are target areas of the body that, when struck, maximize damage to an opponent. Strikes use gross motor skills as opposed to fine motor skills. The target areas of the body are just that—areas. Pinpoint accuracy on a specific nerve is not needed for the strike to be effective.

Hammer Fist

Striking with the hammer fist concentrates power in a small part of the hand, which, when transferred to the target, can have a devastating effect. The striking surface of the hammer fist is the meaty portion of the hand below the little finger. To execute the hammer fist strike, Marines-

Make a fist and bend the arm at approximately a 45- to 90-degree angle. At the same time, rotate the right hip and right shoulder backward.

Thrust the fist forward onto the opponent while rotating the right hip and shoulder forward.

Rotate the wrist so the hammer fist makes contact on the opponent.

Chin Jab

The chin jab can immediately render an opponent unconscious and cause extensive damage to the neck and spine. The striking surface is the heel of the palm of the hand. To execute the chin jab, Marines—

Bend the right wrist back at a 90-degree angle with the palm facing the opponent and the fingers pointing up.

Keep the right arm bent and close to the body. Extend the hand into a concave position with the fingers slightly spread apart.

Step forward with the left foot toward the opponent, keeping the feet approximately shoulder-width apart and the knees bent. This is done to close with the opponent.

Keep the right arm bent and close to the side. Thrust the palm of the hand directly up under the opponent's chin. At the same time, rotate the right hip forward to drive the body weight into the attack to increase the power of the strike. The attack should travel up the centerline of the opponent's chest to his chin.

Knife Hand

The knife hand is one of the most versatile and devastating strikes. The striking surface is the cutting edge of the hand, which is the meaty portion of the hand below the little finger extending to the top of the wrist. The striking surface is narrow, allowing strikes on the neck between the opponent's body armor and helmet. The knife hand strike is executed from one of three angles: outside, inside, and vertical.

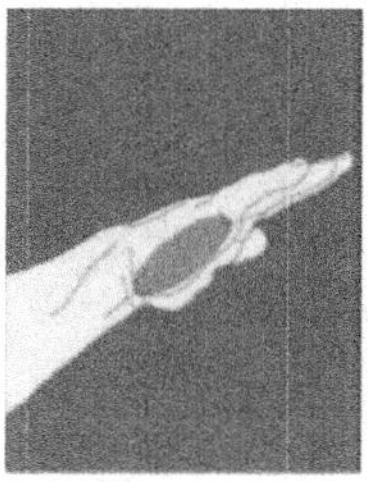

Outside Knife Hand. To execute the outside knife hand strike, Marines—

Execute a knife hand by extending and joining the fingers of the right hand and placing the thumb next to the forefinger (like saluting).

Retract the right hand. At the same time, rotate the right hip and right shoulder backward.

Thrust the knife hand forward (horizontally) onto the opponent while rotating the right hip and shoulder forward.

Inside Knife Hand. To execute the inside knife hand strike, Marines—

Execute a knife hand.

Bring the right hand over the left shoulder. At the same time, rotate the right shoulder forward and the left hip forward.

Thrust the knife hand forward (horizontally) onto the opponent while rotating the right hip and shoulder forward and the left shoulder backward.

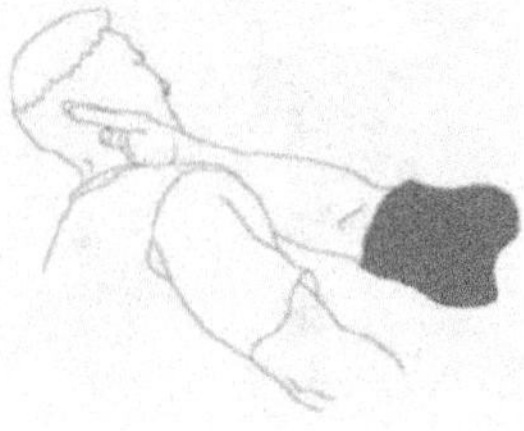

Vertical Knife Hand. When thrown vertically, the knife hand strike comes straight down in a straight line.

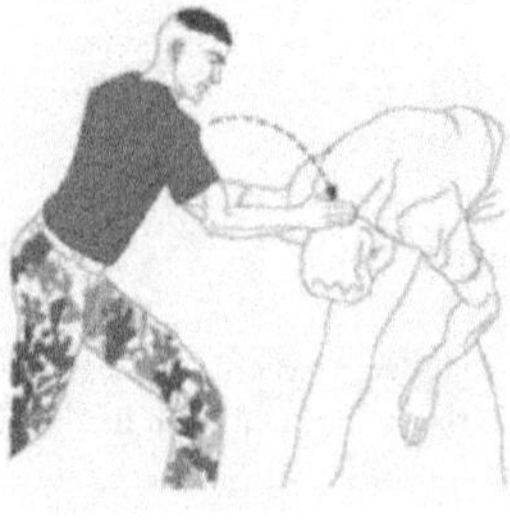

Eye Gouge

The eye gouge is used to attack an opponent's eyes, blinding him so follow-up strikes can be executed. The striking surface is the tips of the fingers or thumb. To execute the eye gouge, Marines—

Extend the right hand with the fingers slightly spread apart to allow entry into the eye sockets.

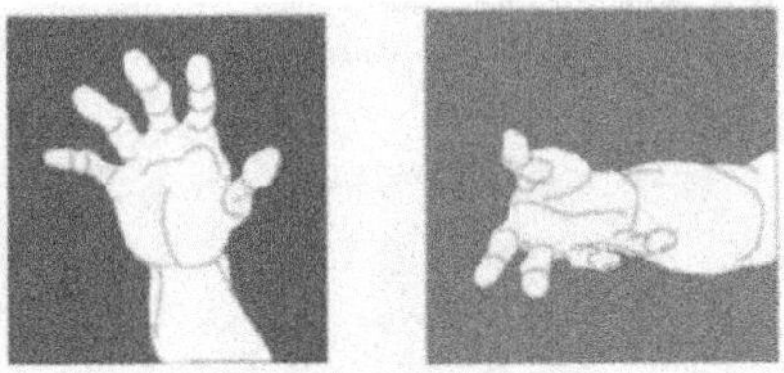

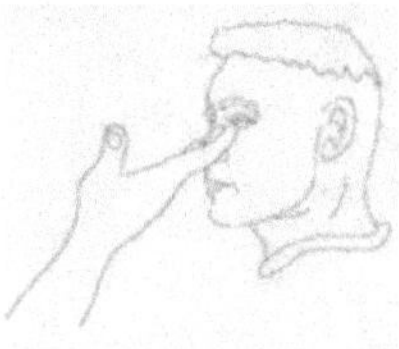

Place the palm of the hand either toward the ground or toward the sky and thrust the right hand forward into the opponent's eyes.

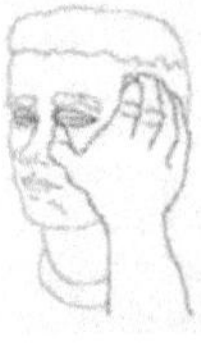

Thrust the hand forward at the opponent's nose level so the fingers or thumb slide naturally into the grooves of the opponent's eye sockets.

Elbow Strikes

The elbow is a powerful weapon that can be used in several different ways to attack virtually any part of an opponent's body. Elbow strikes can be performed either vertically (upward or downward) or horizontally (forward or reverse). The striking surface is 2 inches above or below the point of the elbow, depending upon the angle of attack, the opponent's attack angle, and the position of the opponent.

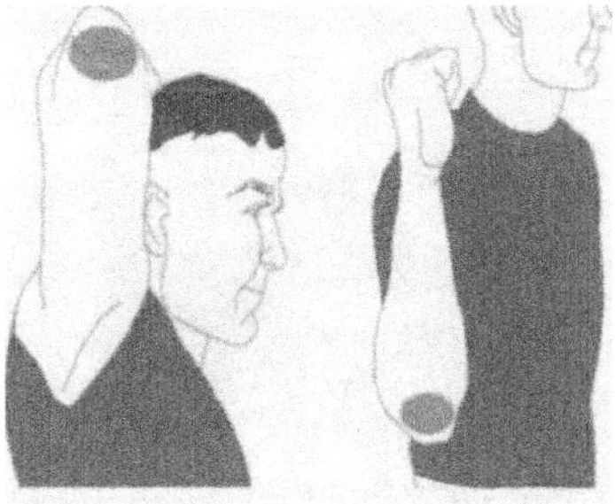

Above Elbow Below Elbow

Vertical Elbow Strike (Up). To execute an upward vertical elbow strike, Marines-

Bend the right elbow, keeping the fist close to the body. The fist is at shoulder level and the elbow is next to the torso.

Thrust the elbow vertically upward toward the opponent while rotating the right shoulder and hip forward to generate additional power.

Contact the opponent with the right forearm 2 inches above the point of the elbow.

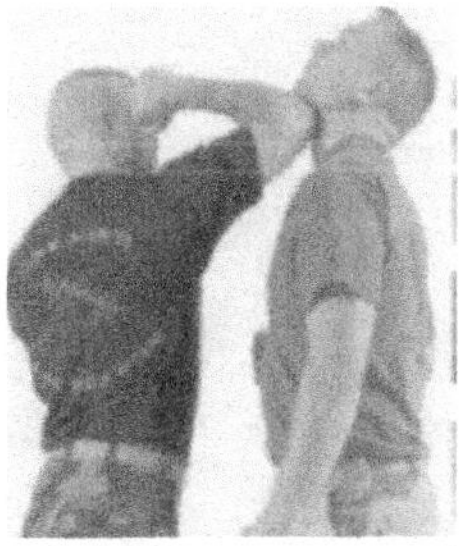

Vertical Elbow Strike (Down). To execute a downward vertical elbow strike, Marines—

Bend the right elbow, keeping the fist close to the body. The fist is on the shoulder and the elbow is raised well above the shoulder.

Thrust the elbow vertically downward toward the opponent while dropping body weight into the attack to generate additional power.

Contact the opponent with the right triceps 2 inches above the point of the elbow.

Horizontal Elbow Strike (Forward). To execute the forward horizontal elbow strike, Marines-

Tuck the right fist near the chest with the palm heel facing the ground.

Thrust the right elbow horizontally forward toward the opponent. The forearm is parallel to the ground.

Rotate the right shoulder and hip forward to generate additional power.

Contact the opponent with the right forearm 2 inches below the point of the elbow.

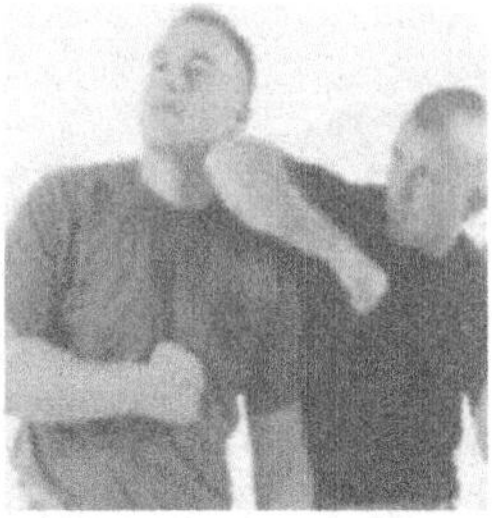

Horizontal Elbow Strike (Rear). To execute the rear horizontal elbow strike, Marines-

Tuck the right fist near the left shoulder with the palm heel facing the ground. At the same time, rotate the right shoulder forward and the left hip forward.

Thrust the right elbow horizontally rearward toward the opponent. The forearm is parallel to the ground and the hand moves toward the direction of the attack.

Rotate the right hip back and the right shoulder backward to generate additional power.

Contact the opponent with the right triceps 2 inches above the point of the elbow.

4. Strikes with the Lower Body

The legs are the body's most powerful weapons because they use the largest muscle groups to generate a strike. Legs are also less prone to injury. The feet are the preferred choice for striking because they are protected by boots. Marines use their feet, heels, and knees to execute kicks, knee strikes, and stomps.

Kicks

The purpose of a kick is to stop an opponent's attack or to create an opening in his defense in order to launch an attack. Kicks can be performed with the left (lead) leg or the right (rear) leg. Kicks with the rear leg have greater power because the hips are rotated into the attack. However, the rear leg is further away from the opponent so a strike with the rear leg will not contact the opponent as quickly as a strike with the lead leg.

Front Kick. The front kick is executed when the opponent is in front of the Marine. The front kick, delivered with either the rear or lead leg, is effective for striking below the waist. Attempting to kick higher results in diminished balance and provides the enemy with a greater opportunity to grab the leg or foot. The striking surfaces arc the toe of the boot or the bootlaces. To execute the front kick, Marines-

Raise the left knee waist high, pivot the hips into the attack, and thrust the left foot forward toward the opponent.

Contact the opponent with the toe of the left boot or bootlaces.

Return to the basic warrior stance.

Side Kick. The side kick, delivered with the lead leg, is effective for striking the knees. The side kick is executed when the opponent is to the side of the Marine. The striking surface is the outside cutting edge of the boot near the heel. To execute the side kick, Marines—

Raise the right knee waist high and rotate the right hip forward.

Thrust the right foot to the right side toward the opponent, turning the foot at a 90-degree angle to maximize the striking surface on the opponent.

Contact the opponent with the cutting edge of the right boot.

Return to the basic warrior stance.

Knee Strikes

Knee strikes are excellent weapons during the close range of close combat fighting. The knee strike is generally delivered in close.

Vertical Knee Strike. The striking surface is the thigh, slightly above the knee. To execute the vertical knee strike, Marines-

Raise the right knee and drive it up forcefully into the opponent. Power is generated by thrusting the leg upward.

Contact the opponent 2 inches above the right knee.

Horizontal Knee Strike. The horizontal knee strike is executed with the leg generally parallel with the ground while rotating the hips to generate power. It is often delivered to the peroneal nerve. The striking surface is the front of the leg, slightly above or below the knee. To execute the horizontal knee strike, Marines-

Raise the right knee, rotate the right hip forward while pivoting on the left foot, and drive the knee horizontally into the opponent.

Contact the opponent 2 inches above the right knee.

Stomps

Stomps are delivered with the feet, usually when the opponent is down. Remember, when the opponent is down, Marines take whatever target is available.

Vertical Stomp. The vertical stomp allows Marines to remain upright and balanced, to rapidly deliver multiple blows with either foot, and to quickly and accurately attack the target. It is the preferred stomp. The striking surface is the flat bottom of the boot or the cutting edge of the heel. To execute the vertical stomp, Marines—

Raise the right knee above the waist with the right leg bent at approximately a 90-degree angle.

Drive the flat bottom of the right boot or the cutting edge of the right heel down onto the opponent forcefully. At the same time, bend the left knee slightly to drop the body weight into the strike.

Ax Stomp. The striking surface of the ax stomp is the cutting edge of the heel. To execute the ax stomp, Marines-

Raise the right heel above the waist, keeping the right leg straight.

Drive the cutting edge of the right heel down onto the opponent forcefully. At the same time, bend the left knee slightly to drop the body weight into the strike.

5. Counters to Strikes

In a close combat situation, an opponent will attempt to strike Marines with punches and kicks. When an opponent uses a strike, Marines must first avoid the strike. This is accomplished by moving quickly and blocking. Next, Marines must get into an offensive position. This allows Marines to use offensive strikes to attack the opponent. Regardless of the strike, the counter to a strike requires Marines to move, block, and strike.

Move

The first step in countering a strike is to move out of the way of the strike's impact. Movement removes Marines from the opponent's intended strike point and positions Marines to attack. Movement is executed at approximately a 45-degree angle to the front or rear. Movement is always initiated from the basic warrior stance. Return to the basic warrior stance with the toe of the lead foot pointing toward the opponent once the movement is complete.

Block

Different blocks are executed based on the strike. These will be covered by individual counters.

Strike

Any of the upper body or lower body strikes can be executed as a follow-on attack or part of the counter to an opponent's strike. The follow-on strike is determined by the angle to the opponent, the position of the opponent, and the opponent's available vulnerable target areas.

Counters to Punches

Counter to a Lead Hand Punch. This counter is used when the opponent throws a lead hand punch. To execute the counter to the lead hand punch, Marines-

Step forward and to the left at approximately a 45-degree angle, moving in to the outside of the opponent's attacking arm.

Raise the left arm and block or deflect the opponent's lead hand with the palm of the hand or the meaty portion of the forearm.

Hit and stick by leaving the left arm against the opponent's right arm while stepping forward and to the left at approximately a 45-degree angle to close with the opponent.

Execute a finishing technique, such as a strike or a kick, to the opponent's exposed target areas.

Counter to a Rear Hand Punch. This counter is used when the opponent throws a rear hand punch. To execute the counter to the rear hand punch, Marines—

Step forward and to the left at approximately a 45-degree angle, moving in to the outside of the opponent's attacking arm.

Raise the left arm and block or deflect the opponent' s rear hand with the palm of the hand or the meaty portion of the forearm.

Hit and stick by leaving the left arm against the opponent's right arm while stepping forward and to the right at approximately a 45-degree angle to close with the opponent.

Execute a finishing technique, such as a strike or a kick, to the opponent's exposed target areas.

Counters to Kicks

Counter to a Front Kick (Left or Lead Leg). This counter is used when the opponent executes a front kick with his left leg. To execute the counter to a front kick, Marines-

Step forward and to the right at approximately a 45-degree angle, moving in to the outside of the opponent's striking leg.

Lower the left arm and block or deflect the opponent' s leg with the palm of the hand or the meaty portion of the forearm.

Hit and stick by leaving the left arm against the opponent's leg while stepping forward and to the left at approximately a 45-degree angle to close with the opponent.

Execute a finishing technique, such as a strike or a kick, to the opponent's exposed target areas.

Counter to a Front Kick (Right or Rear Leg). This counter is used when the opponent executes a front kick with his right leg. To execute the counter to a front kick, Marines—

Step forward and to the left at approximately a 45-degree angle, moving in to the outside of the opponent's striking leg.

Lower the left arm and block or deflect the opponent' s leg with the palm of the hand or the meaty portion of the forearm.

Hit and stick by leaving the left arm against the opponent's leg while stepping forward and to the right at approximately a 45-degree angle to close with the opponent.

CHAPTER 5

THROWS

This chapter describes all techniques for a right-handed person. However, all techniques can be executed from either side.

The Marine is depicted in camouflage utilities. The opponent is depicted without camouflage.

Marines use throwing techniques to maintain the tactical advantage and to throw the opponent to the ground during close combat. Throws apply the principles of balance, leverage, timing, and body position to upset an opponent's balance and to gain control by forcing the opponent to the ground. Throwing techniques are effective because they are size- and gender-neutral, and they rely on the momentum and power generated by the opponent rather than the strength or size of the Marine. Marines also execute a throw as a devastating attack against an opponent, possibly causing unconsciousness or broken limbs. When Marines execute throws, they must maintain balance and, simultaneously, prevent the opponent from countering a throw or escaping after being forced to the ground.

1. Turning Throw

Marines use a turning throw to take the opponent to the ground while they remain standing. A turning throw can also be executed from a stationary position. It is particularly effective if Marines and the opponent are wearing gear. To execute the turning throw, Marines-

Grasp the opponent's right wrist with the left hand.

Step forward with the right foot, place it against the outside of the opponent's right foot, and pivot so the back of the heel is next to the middle of the opponent's foot.

Hook the opponent's right arm with the right arm and pinch his arm between the biceps and forearm, touching the opponent with the body.

Pull the opponent's wrist downward, keeping it close to the body.

Pivot to the left on the ball of the foot and continue pulling downward on the opponent's wrist while rotating the wrist outward to off-balance him.

2. Hip Throw

Marines use a hip throw to take the opponent to the ground while they remain standing. A hip throw is particularly effective if the opponent is moving forward or pushing. Marines use the opponent' s forward momentum to execute the hip throw. To execute the hip throw, Marines-

Grasp the opponent's right wrist with the left hand.

Step forward with the right foot and place it against the outside of the opponent's right foot.

The back of the heel should be next to the opponent' s foot.

Step back with the left foot and rotate on the ball of the foot. The back of the heel is next to the opponent' s toe. The knees are bent.

Rotate at the waist and hook the right hand around the back of the opponent's body (anywhere from his waist to his head). The side and hip should be against the opponent.

Rotate the hip against the opponent. The hips must be lower than the opponent's.

Use the right hand to pull the opponent up on the hip to maximize contact.

Pull the opponent's arm across the body and, at the same time, lift the opponent off the ground slightly by bending at the waist, straightening the legs, and rotating the body to the left.

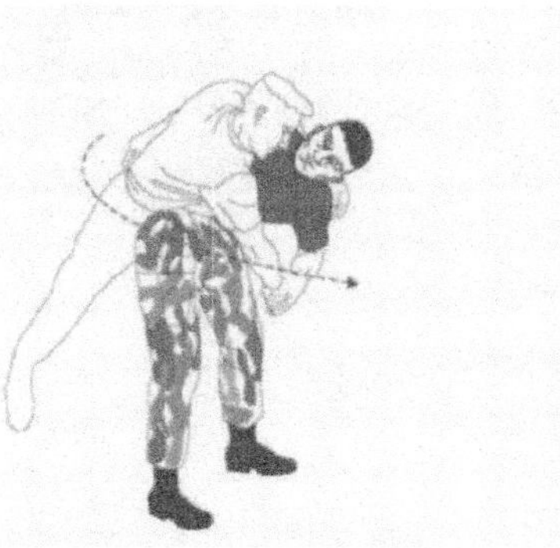

3. Leg Sweep

Marines use a leg sweep to take the opponent to the ground while they remain standing. A leg sweep is particularly effective if the opponent is already off-balanced and moving backward or pulling on the Marine. To execute the leg sweep, Marines—

Grab the opponent's right wrist with the left hand and grab the opponent' s left shoulder with the right hand.

Note: Grab the opponent's clothing or gear if his wrist and shoulder cannot be grabbed.

Step forward with the left foot and place it on the outside of the opponent's right foot.

Pull the opponent's wrist downward, close to the body, and push his shoulder backward to off-balance him.

Raise the right knee no higher than the waist.

Kick the foot past the opponent's right leg.

Use the heel of the boot to make contact with the opponent's calf (anywhere from the top of the calf down to the Achilles tendon or on the inside of the calf).

Sweep the opponent to the ground.

CHAPTER 6

CHOKES AND HOLDS

This chapter describes all techniques for a right-handed person. However, all techniques can be executed from either side.

The Marine is depicted in camouflage utilities. The opponent is depicted without camouflage.

WARNING

During training, never execute a choke at full force or full speed and never hold a choke for more than 5 seconds.

When Marines correctly perform a choke, they render an opponent unconscious in as little as 8 to 13 seconds. Chokes are easily performed regardless of size or gender. Marines must know how to apply chokes and how to counter a choke or a hold executed by an opponent.

1. Types of Chokes

There are two types of chokes: an air choke and a blood choke. An air choke closes off the airway to the lungs, thereby preventing oxygen from reaching the heart. A blood choke cuts off the blood flow to the brain. Both types can result in unconsciousness and eventual death for an opponent.

Air Choke

An air choke is performed on the opponent's windpipe (or trachea), cutting off the air to the lungs and heart. If Marines execute the air choke properly, the opponent loses consciousness within 2 to 3 minutes. Due to the length of time it takes to immobilize the opponent, air chokes are not recommended.

Blood Choke

A blood choke is performed on the opponent's carotid artery, which carries oxygen-enriched blood from the heart to the brain. The carotid artery is located on both sides of the neck. If Marines execute a blood choke properly, the opponent will lose consciousness within 8 to 13 seconds. The blood choke is the preferred choke because its intended effect (i.e., the opponent losing consciousness) can be executed quickly, ending the fight.

2. Chokes

Front Choke

Marines execute a front choke when they are facing the opponent. Marines use the opponent's lapels or collar to execute a front choke. To execute the front choke, Marines-

Grab the opponent's right lapel with the right hand, making certain that the knuckles or the back of the hand are against the opponent's carotid artery on the right side of his neck.

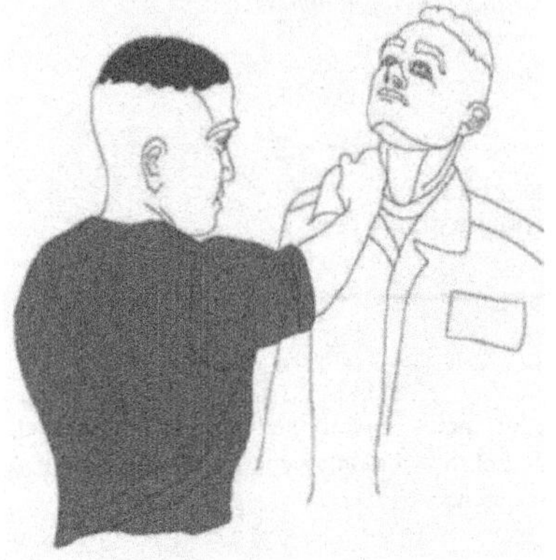

Keep the right hand against the opponent's neck, reach under the opponent's right arm with the left hand, grab the opponent's left lapel, and form an X with the wrists.

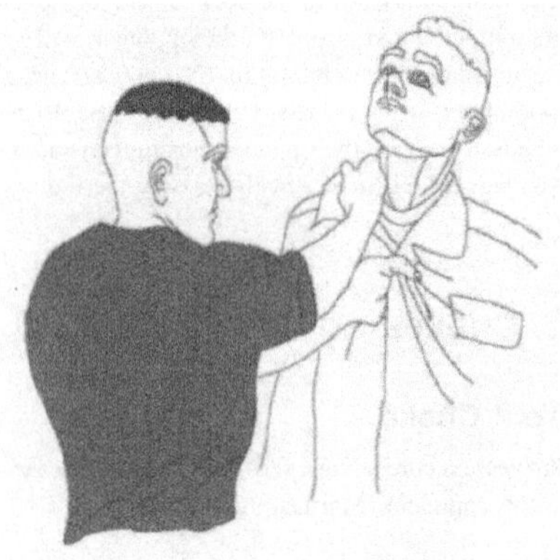

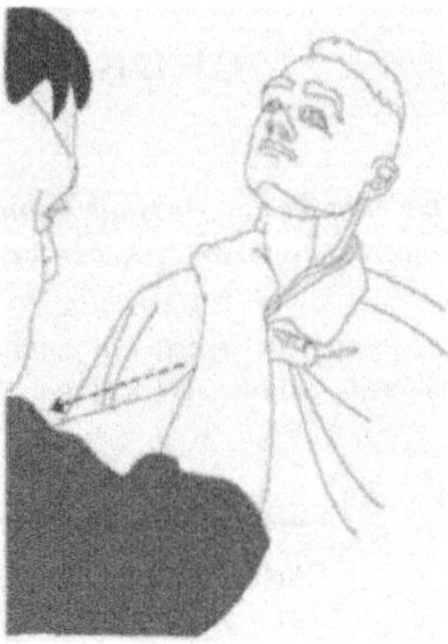

Press the right hand against the opponent's carotid artery.

Pull his lapel to the left with the left hand.

Side Choke

Marines execute a side choke when they are facing the opponent. The side choke is particularly effective when deflecting a punch thrown by an opponent. To execute the side choke, Marines—

Use the left hand to parry the opponent's arm inboard (to the inside of the opponent's reach).

Bring the right arm underneath the opponent's arm and up around the front of his neck.

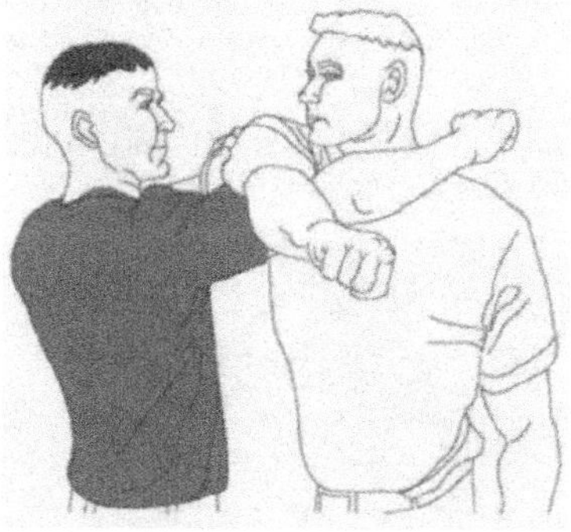

Extend the fingers, place the back of the forearm against the opponent's neck just below his ear, and press the carotid artery on his neck.

Note: If Marines are unable to place their fingers against the opponent's neck, they may use the back of the thumb or wrist.

Reach, with the left hand, around the back of the opponent's neck and clasp the hands together.

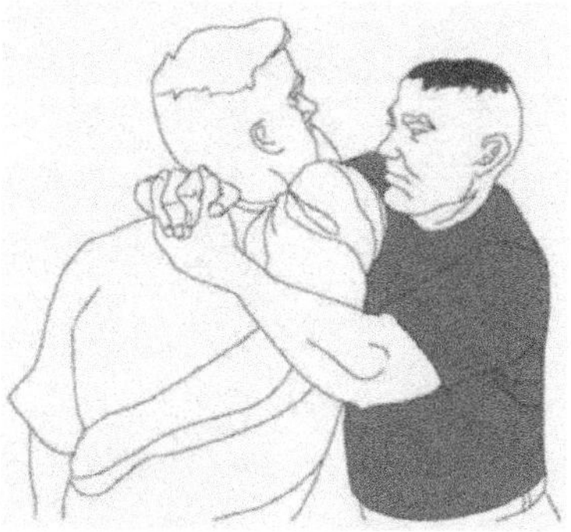

Pull the opponent toward the chest by pulling the clasped hands toward the chest.

Exert pressure on the side of the opponent's neck with the forearm.

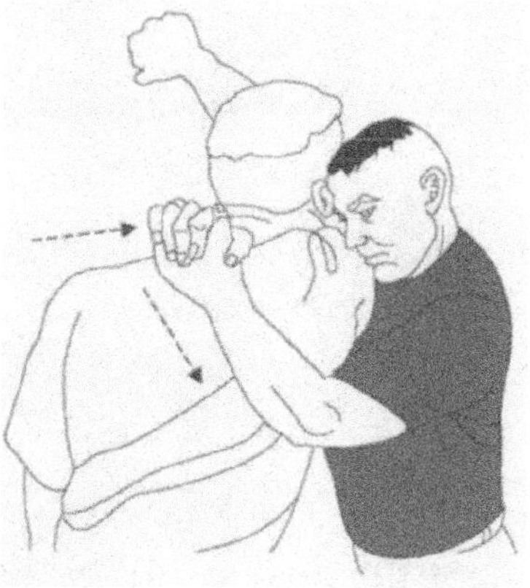

Rear Choke

Marines execute a rear choke when they are behind the opponent, the opponent is on the ground, or when they are taking the opponent to the ground. To execute the rear choke, Marines—

Reach, with the right arm, over the opponent's right shoulder and hook the bend of the arm around his neck.

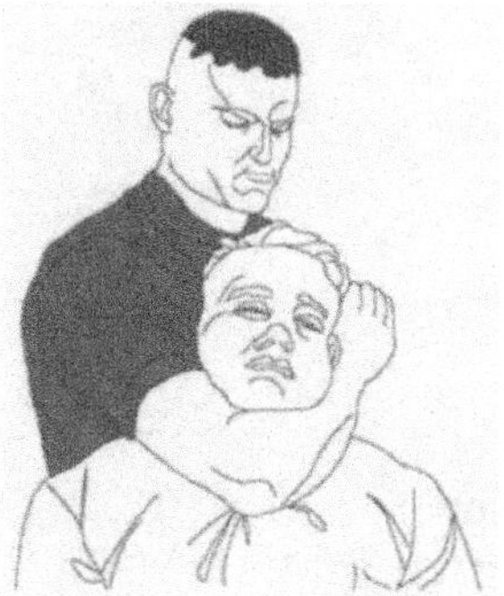

Clasp both hands together.

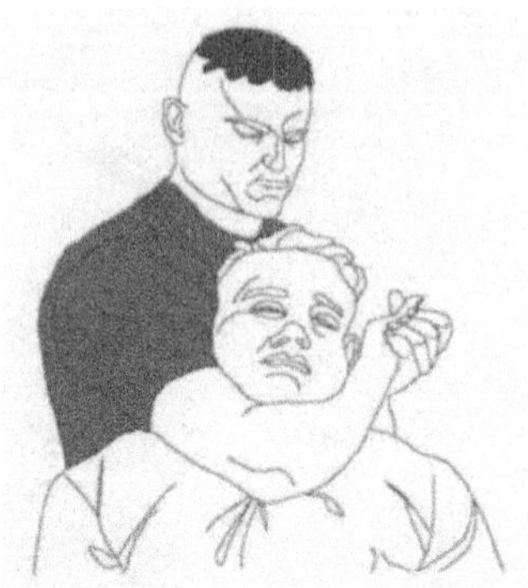

Exert pressure with the biceps and forearms on both sides of the opponent's neck on his carotid arteries.

Maintain pressure with the biceps and forearms on both sides of the neck and draw the opponent closer by drawing the right arm in.

Figure-Four Choke

A variation of the rear choke is the figure-four choke. The figure-four choke allows Marines to gain more leverage than the rear choke. If the rear choke cannot be secured, the figure-four variation is employed to increase the pressure of the choke on the opponent. To execute the figure-four variation of the rear choke, Marines—

Apply a rear choke. The Marine's body should be against the opponent's body.

Grasp the left biceps with the right hand and place the left hand against the back of the opponent's head.

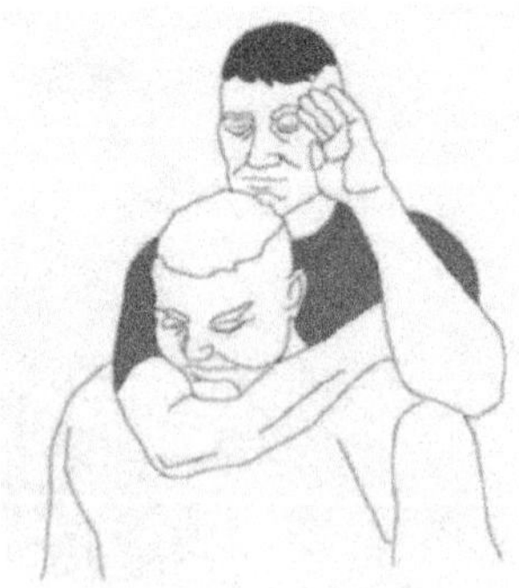

Push the opponent's head forward and down with the left hand.

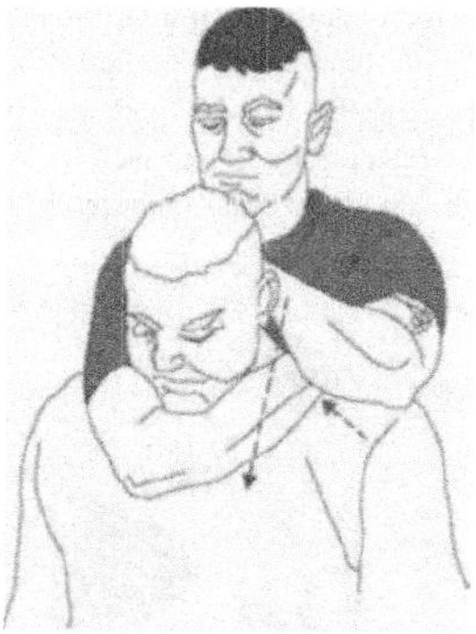

Draw the right arm in, maintaining pressure with the bicep and forearm on both sides of the opponent's neck.

3. Counters to Chokes and Holds

During a close combat situation, an opponent may apply a choke or hold on a Marine. If the opponent correctly applies a choke, a Marine quickly loses consciousness. If a choke is not executed properly, it often results in a hold, typically a bear hug or a headlock. A hold allows the opponent to control a Marine and removes the Marine's ability to attack. It is important for Marines to extract themselves from chokes and holds, regain the tactical advantage, and counter with strikes.

Although a choke causes unconsciousness in 8 to 13 seconds for a blood choke and 2 to 3 minutes for an air choke, the first movement in any counter to a choke is to clear the airway. Marines use softening techniques to loosen an opponent's grip and to clear their airway. Softening techniques are particularly effective if Marines lack the physical strength of their opponent. These techniques include groin strikes, eye gouges, foot stomps, etc. Softening techniques are not offensive; rather, they are used to loosen an opponent's hold.

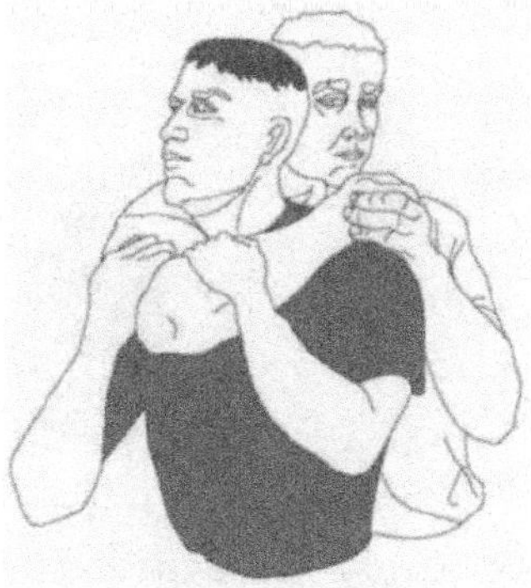

The second movement is to tuck the chin. Once the airway is clear, Marines tuck their chins to prevent the opponent from reapplying the choke.

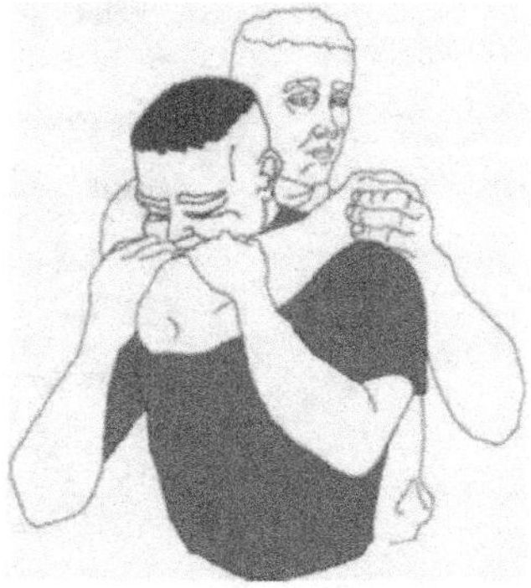

Counter to a Front Choke

Marines use a counter to a front choke when the opponent approaches from the front and uses both hands to choke a Marine around the throat. To execute the counter to the front choke, Marines—

Grasp the opponent's right forearm (where the elbow bends) with the left hand and apply downward pressure on the opponent's radial nerve with the fingers.

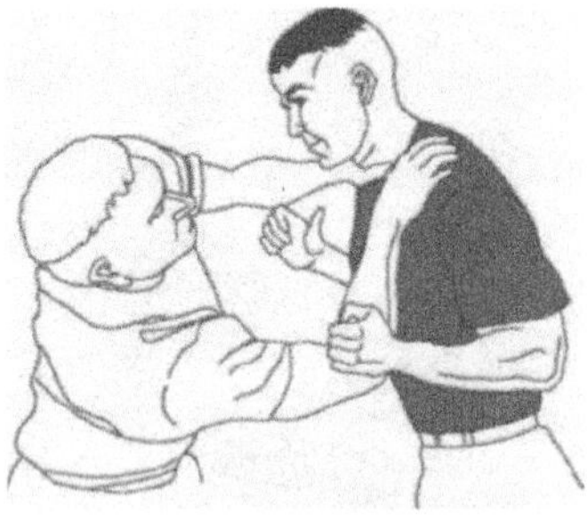

Execute a chin jab to the opponent's chin with the right hand. To generate power into the strike, bring the left foot to the outside of the opponent's right foot and rotate the hips into the strike.

Counter to a Rear Choke

Marines execute a counter to a rear choke when the opponent approaches from the rear and puts his right arm around a Marine's throat. To execute the counter to the rear choke, Marines-

Grasp the opponent's forearm (at the radial nerve) and bicep with both hands and pull down just enough to clear the airway. Once the airway is clear, tuck the chin to protect the airway and to prevent the opponent from reapplying the choke.

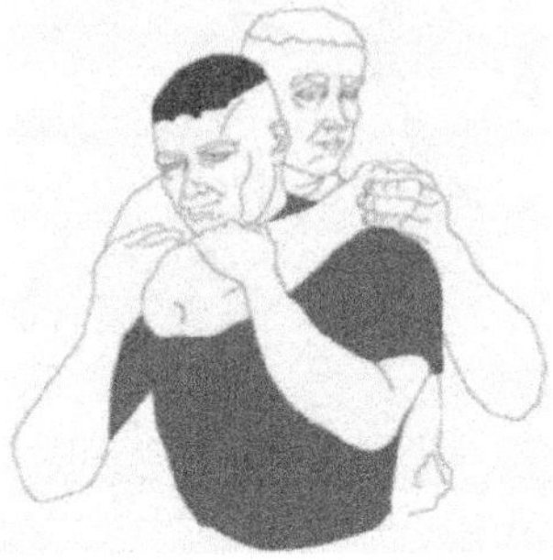

Step behind the opponent's right leg with the left foot, keeping both legs bent (almost in a squatting position).

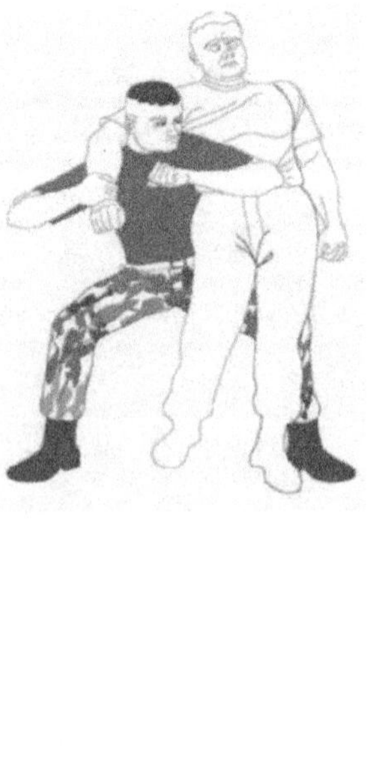

Strike and drive the left elbow into the opponent's torso while rotating the hips and pivoting to the left, throwing the opponent back and to the ground.

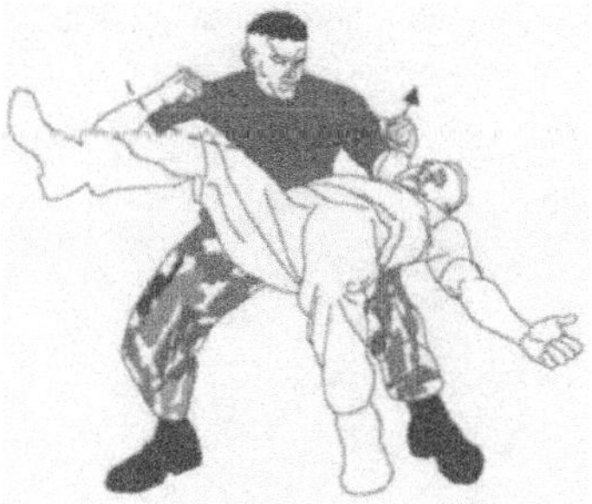

Counter to a Front Headlock

Marines use a counter to a front headlock when the opponent approaches from the front and puts his right arm around the Marine's neck, bends the Marine forward, and locks the Marine's head against his hip. To execute the counter to a front headlock, Marines—

Grasp the opponent's wrist and forearm with both hands and pull down to clear the airway. Maintain control of the opponent's wrist throughout the move. Once the airway is clear, tuck the chin to protect the airway and to prevent the opponent from reapplying the choke.

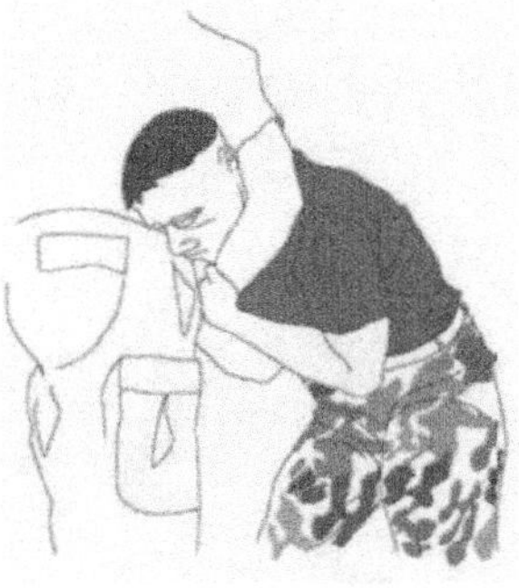

Move the right hand and arm across the opponent's torso.

Step forward and to the left with the left foot at a 45-degree angle.

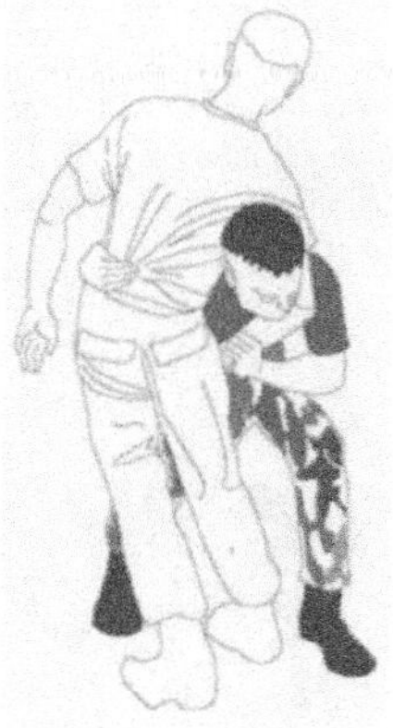

Execute a sweep with the right foot against the opponent's right leg. At the same time, push against the opponent's chest with the right arm and shoulder to generate power in the sweep.

Counter to a Rear Headlock

Marines use a counter to a rear headlock when the opponent approaches from the rear and puts his right arm around the Marine's neck, bends the Marine forward, and locks the Marine's head against his hip. To execute the counter to the rear headlock, Marines-

Grasp the opponent's wrist and forearm with the right hand and pull down to clear the airway. Once the airway is clear, tuck the chin to protect the airway and to prevent the opponent from re-applying the choke.

Reach over the opponent's right shoulder with the left arm.

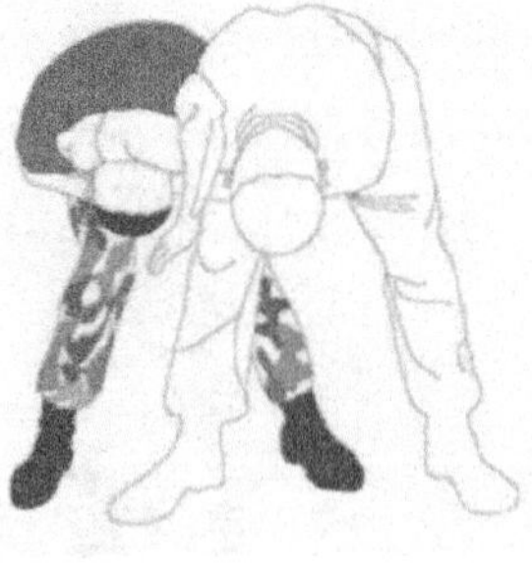

Grab any part of the opponent's face (chin, nose, eyes) and pull back while rising to a standing position.

Execute, with the right hand, a hammer fist strike to the opponent's exposed throat.

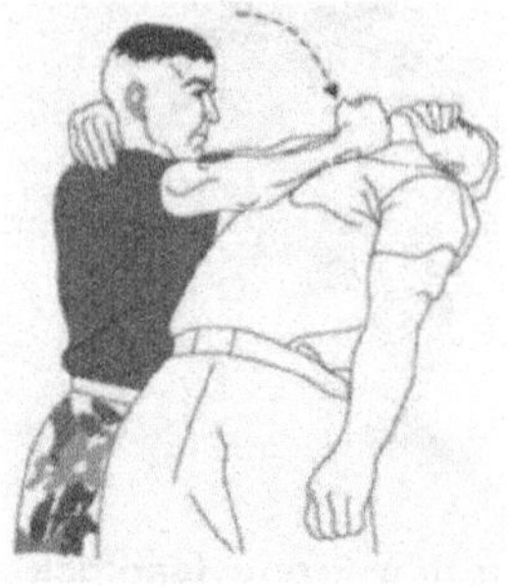

Counter to a Front Bear Hug

Marines execute a counter to a front bear hug when the opponent approaches from the front and puts both of his arms around the Marine's body, trapping the Marine's arms to the sides. To execute the counter to a front bear hug, Marines —

Step forward and to the left with the left foot at a 45-degree angle to the outside of the opponent's right leg, keeping the left leg bent.

Grasp the opponent's torso or arms to gain balance and to assist in throwing the opponent. It may be helpful to hook the opponent's right arm with the left arm.

Drive the right arm and shoulder forward and, at the same time, bring the right leg forward and sweep the opponent's right leg, bringing him to the ground.

Counter to a Rear Bear Hug

Marines use a counter to a rear bear hug when the opponent approaches from the rear and puts both of his arms around the Marine's body, trapping the Marine's arms to the sides. To execute the counter to a rear bear hug, Marines—

Step behind the opponent's right leg with the left foot, keeping both legs bent (almost in a squatting position). The left side of the body should be against the opponent's.

Pivot the hip, turning the body to the left and throwing the opponent back over the bent leg.

CHAPTER 7

GROUND FIGHTING

This chapter describes all techniques for a right-handed person. However, all techniques can be executed from either side.

In drawings, the Marine is depicted in woodland camouflage utilities; the opponent is depicted witlwut camouflage. In photographs, the Marine is depicted in woodland camouflage utilities; the opponent is depicted in desert camouflage utilities.

Marines should avoid being on the ground during a close combat situation because the battlefield may be covered with debris and there is an increased risk of injury. However, many close combat situations involve fighting on the ground. The priority in a ground fight is for Marines to get back on their feet as quickly as possible. In any ground fighting scenario, Marines will usually end up in one of four positions with the opponent. The offensive positions, in which Marines have a tactical advantage, are the guard and mount. The defensive positions, which are used as counters when the opponent has the tactical advantage, are the counter to the guard and the counter to the mount. Marines can also employ chokes during ground fighting to quickly end a fight.

1. Offensive Ground Fighting

Guard Position

Marines execute a guard position when the opponent is on top and the legs are wrapped around the opponent's legs. If the opponent is on top, he may try to choke the Marine, but the Marine still has the tactical advantage because the Marine is in a position to control the situation. To execute the guard position technique, Marines—

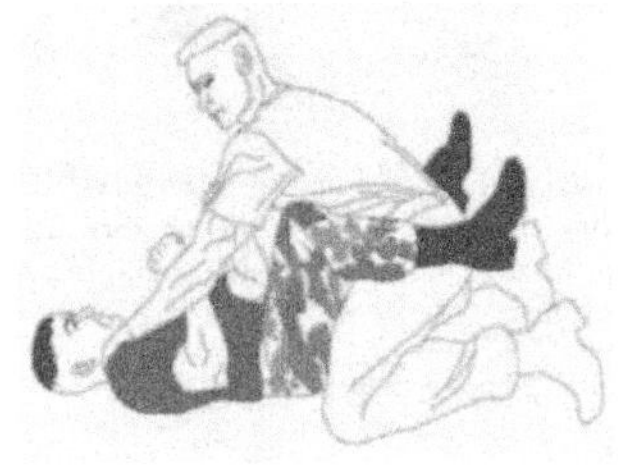

Trap the opponent's hands on the chest by crossing the hands on the chest so the forearms are resting on the opponent's forearms. Apply pressure with the elbows or forearms.

Strike the outside of the opponent's thigh with the cutting edge of the right heel. This causes the opponent to jerk to that side.

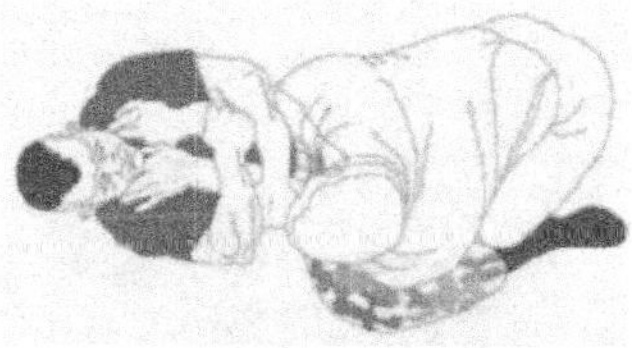

Move the head quickly to the left and swivel the hips to the right. At the same time, bring up both of the legs. Both legs are on the right side of the opponent's body.

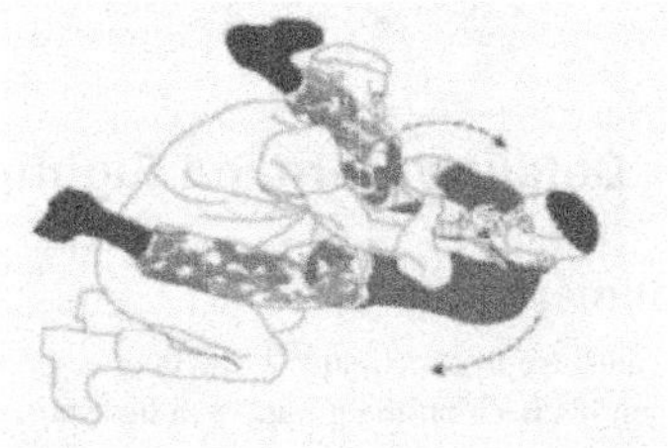

Bring the right leg down, hooking the opponent's neck and head. Exert downward pressure to roll him over on his back. Grasp and maintain control of the opponent's left arm. Upon completion of the move, the Marine is sitting up with legs bent over the opponent while maintaining control of his arm.

Keep legs and knees bent. Maintain pressure against the opponent's neck with the back of the right foot and against his side with the left foot underneath his armpit. Squeeze the knees together, locking the opponent's arm.

Pull the opponent's arm straight up and fall back sharply, pulling his arm to the side in the direction of his little finger.

Return to standing.

Mount Position

Marines execute the mount if the opponent is lying on his back on the ground and the Marine is on top with legs wrapped around the opponent's body. This position is an offensive position because Marines are in a better position to control the opponent and to execute ground fighting techniques. To execute the mount position technique, Marines—

Grab the opponent's wrists or forearms with the hands. Hold them tightly against the torso.

Maintain control of the opponent's arms with the left arm and apply pressure to the opponent's brachial plexus (tie in) with the right hand.

Bring the left foot close to the opponent's right armpit and swing the right leg across the opponent' s head.

Keep the legs and knees bent. Maintain pressure against the opponent's neck with the back of the right foot and against his side with the left foot which is underneath his armpit. Squeeze the knees together, locking the opponent's arm.

Pull the opponent's arm straight up and fall back sharply, pulling his arm to the side in the direction of his little finger.

Return to standing.

2. Defensive Ground Fighting

Counter to the Guard

Marines use this technique if the opponent is lying on his back on the ground and the Marine is kneeling on the ground between the opponent's legs. To execute the counter to the guard position, Marines—

Strike the opponent's femoral nerve, located on the inside of the thigh, with the elbows. This forces the opponent to separate his legs.

Strike the opponent's groin with the right fist.

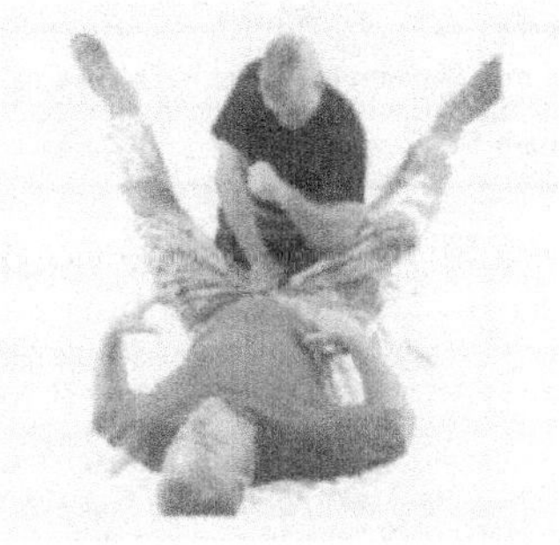

Hook the left arm underneath the opponent's right knee from the inside. Throw the opponent's leg over the head.

Duck the head quickly and move to the left.

Return to standing.

Counter to the Mount

Marines use this technique when they are lying on their back on the ground, the opponent is mounted on top of the Marine, and the opponent's legs are wrapped around the Marine. The opponent has the tactical advantage. To execute the counter to the mount position, Marines—

Grab the opponent's gear or clothing on his upper torso and pull him down close.

Use the right arm to hook the opponent's left arm, from the inside around the outside, above his elbow. With the right foot, hook the opponent's left leg or ankle.

Draw the elbow in to bend the opponent's elbow, bringing him down close. The arm must be hooked above the opponent's elbow in order to bend it.

Strike the opponent's side with the left hand.

Push the opponent over and roll him off to the right side.

Maintain control of the opponent's hooked arm and move to a kneeling position.

Move to a standing position while maintaining pressure on the opponent's arm and using the knee to apply pressure against the opponent's elbow.

3. Ground Fighting Chokes

WARNING

During training, never execute a choke at full force or full speed. Never hold a choke for more than 5 seconds.

The priority in ground fighting is for Marines to get back on their feet as quickly as possible. Sometimes, Marines can quickly end a ground fight by executing a choke on the opponent. When performed correctly, a choke can render an opponent unconscious in as little as 8 to 13 seconds. Chokes are easily performed regardless of size or gender. The chokes performed during ground fighting are the same as those performed while standing (see chap. 6).

Ground Fighting Front Choke

The ground fighting front choke is a blood choke performed most effectively from the mount position. The front choke employs the opponent's lapels or collar to execute the choke. To execute the ground fighting front choke, Marines—

Grab the opponent's right lapel with the right hand, making certain that the knuckles or the back of the hand are against the carotid artery on the right side of the opponent's neck.

Keep the right hand pressed against the opponent's carotid artery, reach under the right arm with the left hand, and grab the opponent's left lapel, forming an X with the wrists.

Keep the right hand pressed against the opponent's carotid artery and pull the opponent's left lapel to the left with the left hand.

Ground Fighting Side Choke

The ground fighting side choke is a blood choke performed from the mount position. The ground fighting side choke is particularly effective when the opponent raises his arms and places them on the Marine's chest or throat. To execute the ground fighting side choke, Marines-

Use the left hand to parry the opponent's right arm inboard (to the inside of the opponent's reach). Bring the right arm underneath the opponent's arm and up around the front of his neck.

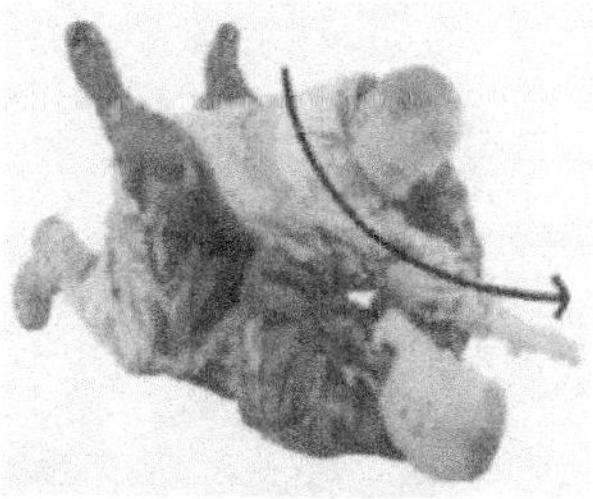

Keep the fingers extended, place the back of the forearm against the opponent's neck just below his ear and press the carotid artery on his neck. Depending on the Marine's or the opponent's body configuration, the back of the thumb or wrist may press against the opponent's neck.

Use the left hand to reach around the back of the opponent's neck and clasp the hands together.

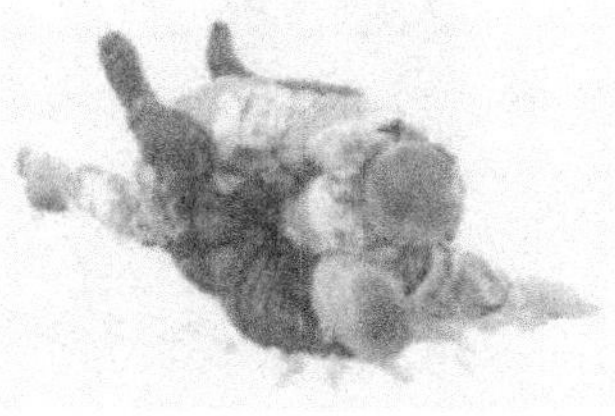

Pull the opponent toward the chest. Use the fore-arm to exert pressure on the side of his neck. This is done by pulling the clasped hands toward the chest.

Ground Fighting Rear Choke

The ground fighting rear choke is a blood choke performed when Marines are behind the opponent. To execute the ground fighting rear choke, Marines—

Wrap the legs around the opponent with the left foot against the inside of his left thigh and the right foot against the inside of his right thigh.

Use the right arm to reach over the opponent's right shoulder and hook the bend of the arm around his neck.

Use the left hand to clasp both hands together.

Exert pressure with the biceps and forearms on both sides of the opponent's neck. This applies pressure to the opponent's carotid arteries. While maintaining pressure with the biceps and fore-arms, draw the opponent closer by drawing the right arm in. To increase the effectiveness of the choke, lean backward by arching the back and pulling the opponent back. At the same time, push the feet against the opponent's thighs.

Figure-Four Choke

A variation of the ground fighting rear choke is the figure-four choke. The figure-four choke allows Marines to gain more leverage on the rear choke. If the rear choke cannot be secured, the figure-four variation may be applied to increase the pressure of the choke on the opponent. To execute the figure-four variation of the rear choke, Marines—

Apply the ground fighting rear choke. The chest should be against the opponent's back.

Grasp the left biceps with the right hand and place the left hand against the back of the opponent's head.

Use the left hand to push the opponent's head forward and down.

Draw the right arm in while maintaining pressure with the biceps and forearm on both sides of the opponent's neck.

Increase the effectiveness of the choke by arching backward and pushing the feet against the opponent's thighs. Continue to pull the opponent back with the right arm while exerting pressure forward with the left hand against the opponent's head.

CHAPTER 8

NONLETHAL TECHNIQUES

This chapter describes all techniques for a riglit-handed person. However, all techniques can be executed from either side.

In drawings, the Marine is depicted in woodland camouflage utilities; the opponent is depicted without camouflage. In photographs, the Marine is depicted in woodland camouflage utilities; the opponent is depicted in desert camouflage utilities.

The Marine Corps' involvement in military operations other than war—e.g., humanitarian, peacekeeping, or evacuation missions—has greatly increased. These missions require skills that span the spectrum of conflict and support operations within a continuum of force. But the Corps' day-to-day existence also demands a responsible use of force. Nonlethal techniques are among the skills Marines use to apply a responsible use of force.

1. Unarmed Restraints and Manipulation

Marines operate within a continuum of force, particularly in support of peacekeeping or humanitarian types of missions. In these situations, Marines must act responsibly to handle a situation without resorting to deadly force. Unarmed restraints and manipulation techniques including joint manipulation, come-alongs, and takedowns can be used to control a subject without resorting to deadly force. Marines must train to become proficient in nonlethal techniques and to respond in a responsible manner. These techniques are referred to as compliance techniques, and they are applied in the third level in the continuum of force.

WARNING

During training, never apply the techniques for unarmed restraints and manipulation at full force or full speed. Use a slow and steady pressure to avoid injury.

Compliance Techniques

Compliance techniques are unarmed restraint and manipulation techniques used to physically force a subject or opponent to comply. Compliance can be achieved through the close combat techniques of-

Pain compliance using joint manipulation and pressure points. (Pain compliance is the initiation of pain to get compliance on the part of the subject.)
Come-along holds.

Principles of Joint Manipulation

Joint manipulation is used to initiate pain compliance and gain control of a subject. It involves the application of pressure on thejoints (elbow, wrist, shoulder, knee, ankle, and fingers). Pressure is applied in two ways:

In the direction in which the joint will not bend. For example, joints such as the knees and elbows only bend in one direction and when pressure is applied in the opposite direction, pain compliance can be achieved. Beyond the point where the joint stops naturally in its range of movement (i.e., it no longer bends).

Since each joint has a breaking point, Marines should apply slow steady pressure only until pain compliance is reached. Continued pressure will break the joint and may escalate the violence of the situation.

Joint manipulation also uses the principle of off-balancing. A subject can be better controlled when he is knocked off balance.

Wristlocks

A wristlock is a joint manipulation that can be applied in a number of ways to achieve pain compliance. The wrist rotates in a number of directions and will bend in a single direction until its movement stops naturally. In a wristlock, pressure is exerted beyond that point by bending or twisting the joint. A wristlock is executed when an opponent tries to grab Marines or is successful in grabbing Marines or their equipment. A wristlock can also be performed by Marines if they wish to initiate control of an opponent.

Basic Wristlock. A basic wristlock is executed when Marines grab the opponent's left hand with the right hand. To execute the basic wristlock, Marines—

Use the right hand to grab the opponent's left hand by placing the thumb on the back of the opponent's hand so that the Marine's knuckles are facing to the left.

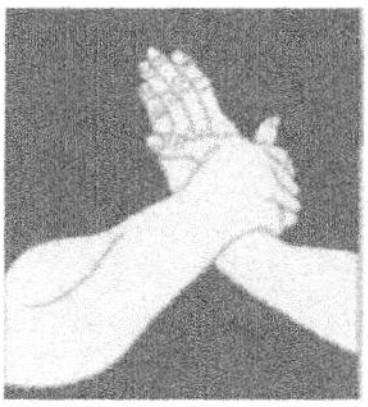

Hook the fingers across the fleshy part of the opponent's palm below the thumb. The fingers are used to anchor the hand so leverage can be applied to twist and bend the joint.

Exert downward pressure with the thumb to bend the opponent'sjoint. Rotate the opponent's hand to the right to twist the joint.

Use the left hand to further control the opponent.

Step in to the opponent to keep the opponent's hand in close to the body to control him and provide more leverage on the wristlock.

Note: When executing the basic wristlock with the left hand, the Marine grabs the opponent's hand so that the Marine's knuckles are facing to the right, and then rotates and twists the opponent's hand to the left.

Reverse Wristlock. A reverse wristlock is executed when Marines grab the opponent's right hand with the right hand. To execute the reverse wristlock, Marines—

Place the right palm on the back of the opponent's right hand and wrap the fingers across the fleshy part of his palm below his little finger.

Twist the opponent's hand to the right while stepping in to place his hand against the chest. Apply downward pressure on the opponent's hand against the chest. Leave the opponent's hand on the chest to fully control the subject and to gain leverage.

Lean forward to use body weight to add additional pressure to the joint.

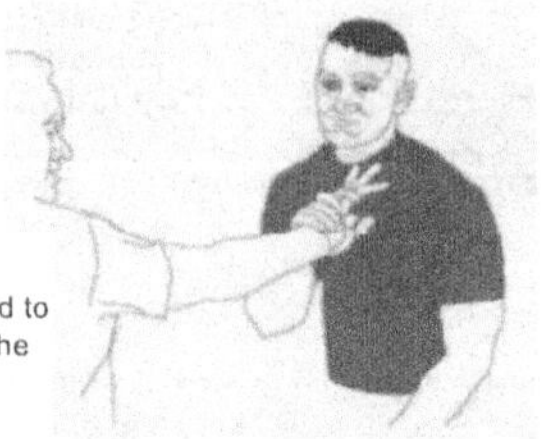

Use the left hand to further control the opponent.

Two-Handed Wristlock. Both hands can be used in the wristlock to maximize the leverage and pressure needed to bend and twist the joint. To execute the two-handed wristlock, Marines—

Place both thumbs on the back of the opponent's hand, thumbs crossed.

Hook the fingers of both hands around the fleshy part of the opponent's palm on both sides of his hand.

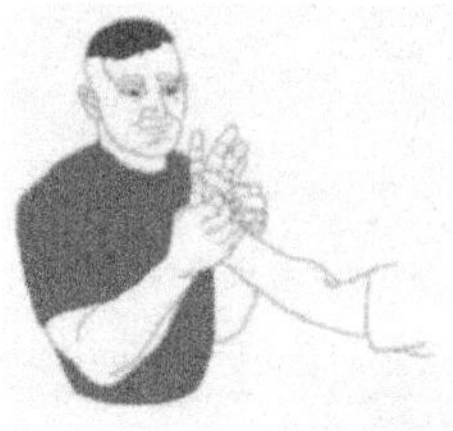

Step into the opponent and apply pressure downward on the back of his hand to bend the joint and rotate his wrist away from the body to twist the joint.

Enhanced Pain Compliance on Wristlock. Enhanced pain compliance techniques are applied in the third and fourth levels in the continuum of force. Additional pain be applied to a wristlock by-

Adding downward pressure to the elbow with the other hand or elbow by using the fingers to pull in on the opponent's radial nerve located on the inside of the forearm. When pressure is added to the opponent's radial nerve, his direction can be controlled.

Applying pressure against the opponent's finger joint to bend it in a direction it cannot bend (i.e., splitting the fingers).

Come-Along Holds

Marines use a come-along hold to control and move an opponent.

Escort Position. A common come-along hold is the escort position. To execute the escort position, Marines—

Face the opponent. Use the left foot to step forward at a 45-degree angle. Turn to face the right side of the opponent.

Use the right hand to firmly grasp the opponent's right wrist. With the left hand, firmly grasp the opponent's right triceps.

Position the opponent's controlled arm diagonally across the torso, keeping his wrist against the right hip. The Marine should be standing to the right of and behind the opponent.

Note: This technique works well when escorting an opponent on either the right or left side. Take caution when escorting an opponent by ensuring his controlled hand is not in a position to grab the holstered weapon. The preferred escort position is from the Marine's left side, so that the opponent is kept further away from the weapon.

Wristlock Come-Along. To execute the wristlock come-along, Marines-

Use the left hand to execute a basic wristlock. Incorporate the right hand in a two-handed wristlock for more control.

Maintain pressure on the opponent's wrist with the right hand, step forward, and pivot around to stand next to the opponent.

Release the left hand, quickly reach under the opponent's arm from behind, and grab his hand.

Use the left hand and apply downward pressure on the opponent's wrist.

Controlling Technique. The following controlling technique is used when an opponent grabs the Marine's wrist. To execute the technique, Marines—

Trap the opponent's hand with the palm of the other hand.

Rotate the opponent's trapped hand up and on his forearm while maintaining downward pressure on his trapped hand.

Apply downward pressure with both hands until the opponent is taken to the ground.

Armbars

An armbar is a joint manipulation in which pressure is applied on a locked elbow, just above the joint, in the direction the joint will not bend. An armbar has to be locked in quickly, but still requires a slow, steady pressure to gain compliance.

Basic Armbar. To execute a basic armbar, Marines-

Use the right hand to grab the opponent's right wrist.

Bring the left hand down on or above the opponent's elbow joint. To gain additional leverage, pivot to face the opponent.

Use the left hand to apply downward pressure on or above the opponent's elbow joint while pulling up on his wrist.

Armbar from a Wristlock. To execute an armbar from a wristlock, Marines-

Use the right hand to grab the opponent's right hand and execute a reverse wristlock.

Bring the left hand down on or above the opponent's elbow joint.

Use the left hand to apply downward pressure on or above the opponent's elbow joint while pulling up on his wrist.

Takedowns

A takedown is used to bring an opponent to the ground to further control him.

Takedown From a Wristlock Come-Along. To take the opponent to the ground from a wristlock come-along, Marines—

Use the right foot to push down on the opponent's calf or Achilles tendon.

Maintain control of the opponent's wrist and elbow and apply a slow, steady pressure to bring him to the ground.

Armbar to a Takedown. This technique is used to take a noncompliant opponent to the ground from an armbar. To execute the armbar takedown, Marines—

Use the right hand to execute a reverse wristlock.

Bring the left hand or forearm down on or above the opponent's elbow joint.

Pivot so the back is facing the opponent and, at the same time, lift the left elbow and slide the body so it is against the opponent, placing the armpit high above the opponent's elbow joint.

Lean back, placing the body weight on the opponent's arm until he complies or is taken to the ground.

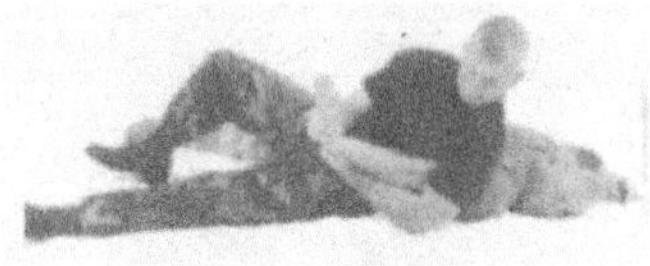

Note: This technique may break the opponent's arm. Therefore, this technique should not be employed if the objective is a nonlethal takedown.

Wristlock Takedown. This technique is used to take a noncompliant opponent to the ground from a basic wristlock and to put him in a position where he can be handcuffed, if necessary. To execute the takedown, Marines-

Use the right hand to execute a basic wristlock.

Incorporate the left hand in a two-handed wristlock.

Apply downward pressure on the wristlock, pivot on the ball of the right foot, and quickly turn to the right to take the opponent to the ground.

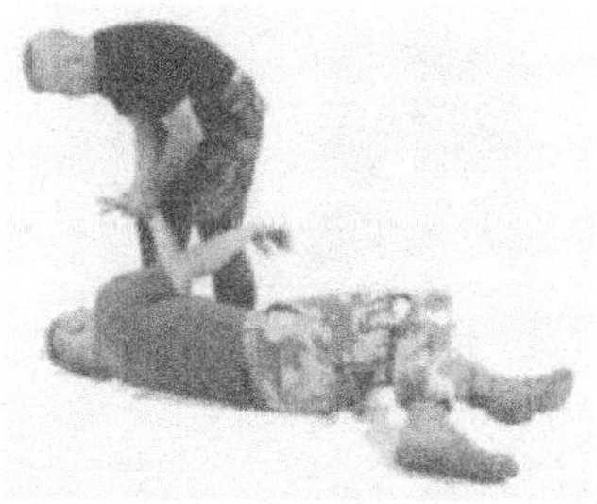

Continue to apply pressure on the wristjoint as the opponent lands on his back with his arm straight in the air.

Slide the left foot under the opponent's back, underneath his armpit.

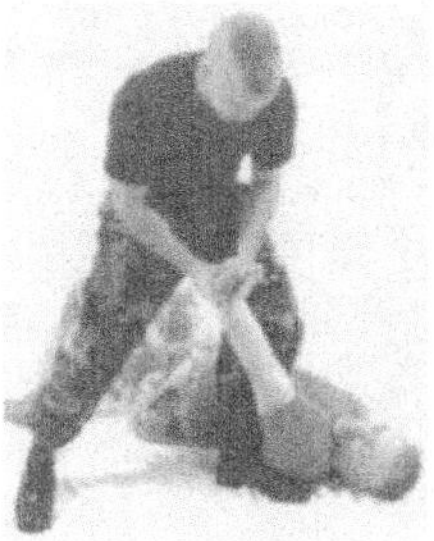

Apply pressure with the knee against the opponent's triceps while pulling back on his arm.

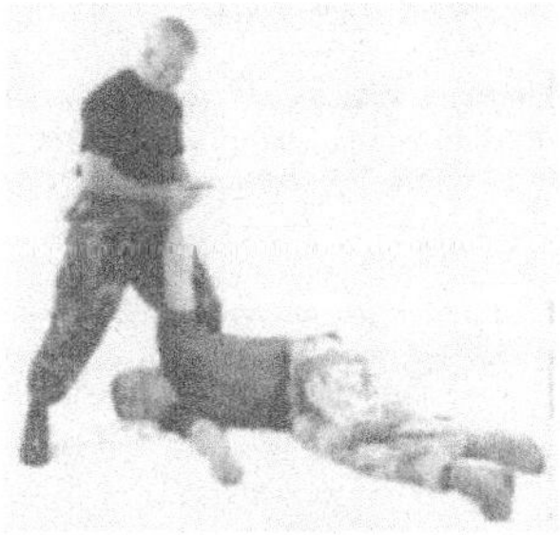

Continue applying pressure with the knee on the opponent's arm. Pivot and step around the opponent's arm to roll him on his stomach.

Kneel down with one knee on the opponent's back. The other knee is placed on the opponent's neck and shoulder on either side of his arm. Apply inward pressure with the knees to lock his arm in place.

Tell the opponent to put his other hand in the middle of his back. Bring the opponent's controlled hand to the center of his back.

Escort Position Takedown. This technique can be used to control a noncompliant opponent from the escort position. To execute the escort position takedown, Marines—

Lock the opponent's arm straight across the body while rotating his wrist away from the body.

Use the left hand or forearm and apply downward pressure above the opponent's elbow where the triceps meet.

Step back with the right foot and, keeping the opponent's hand controlled against the hip, pivot to the right while continuing to apply downward pressure on his arm to bring him to the ground.

Note: This technique works well when escorting a subject on the Marine's right or left side. When taking down a subject from the right side, step back and pivot to the left.

2. Nonlethal Baton

A baton or nightstick can be an effective compliance tool when used correctly. Batons can be used defensively (blocking), offensively (striking), and as a restraining device when needed. In the fourth level of the continuum of force (assaultive [bodily harm]), defensive tactics include baton or nightstick blocks and blows. Blows to the head or other bony parts of the body are considered deadly force. When deadly force is not authorized, Marines must be able to employ blocks, strikes, and restraints with the baton with the minimum of force.

Grips

One-Handed Grip. To execute the one-handed grip, Marines—

Use the right hand to grasp the lower end of the baton, about 2 inches from the end.

Wrap the thumb and index finger around the baton so they are touching one another. The grip on the baton should be firm, but natural.

Two-Handed Grip. To execute the two-handed grip, Marines-

Use the right hand to grasp the lower end of the baton, about 2 inches from the end. Wrap the thumb and index finger around the baton so they touch one another.

Use the left hand to grasp the upper end of the baton, palm down, about 2 inches from the end. The hands should be approximately 10 to 12 inches apart.

Stance and Method of Carry

The basic warrior stance serves as the foundation for initiating nonlethal baton techniques. The method of cany provides effective defensive positions with a wide range of options to control a combative opponent.

One-Handed Carry. To execute the one-handed carry, Marines—

Grip the baton using the one-handed grip.

Elevate the baton, with the gripping hand at a level between the belt and shoulder.

Keep the left hand in the position of the basic warrior stance.

Two-Handed Carry. This carry is effective for blocks. To execute the two-handed carry, Marines-

Grip the baton using the two-handed grip.

Elevate the baton, with the left hand higher than the right hand.

Orient the weapon toward the subject.

Movement

In a nonlethal confrontation, movement may be made to create distance between Marines and the opponent or to close the gap to control him. When facing an opponent, Marines move in a 45-degree angle to either side of the opponent. Moving at a 45-degree angle is the best way to both avoid an opponent's strike and to put Marines in the best position to control the opponent.

Blocking Techniques

One-handed blocks are used when carrying the baton in the one-handed carry. The same one-handed blocks used in combative stick techniques (see page 3-6) apply in nonlethal baton engagements. The blocks include blocks for a vertical strike, a forward strike, and a reverse strike. Because the baton is often carried with two hands, there are also two-handed blocks that are effectively used from this carry. Two-handed blocks are discussed in the following subparagraphs.

High Block. Marines execute a high block to deter a downward vertical attack directed at the head and shoulders. To execute the high block, Marines—

Raise the baton up to a level even with or above the head. The baton should be in a horizontal position to block the blow. The fingers of the left hand should be open and behind the baton.

Place the baton perpendicular to the opponent's striking surface to absorb the impact of the blow.

Bend the elbows slightly to help absorb the impact of the blow. The arms should give with the strike of the blow.

Low Block. Marines execute a low block to deter an upward vertical attack directed at the abdomen, groin, or torso. The opponent's blow can be delivered by a foot, knee, or fist. To execute the low block, Marines—

Lower the baton to a level even with or below the groin. The baton should be in a horizontal position to block the blow. The fingers of the left hand should be open and behind the baton.

Place the baton perpendicular to the opponent's striking surface to absorb the impact of the blow.

Bend the elbows slightly to help absorb the impact of the blow. The arms should give with the strike of the blow.

Right Block. Marines execute a right block to deter an attack directed at the head, neck, flank, or hip. The opponent's blow can be delivered by a foot, knee, fist, or elbow. To execute the right block, Marines-

Thrust the baton in a vertical position to the right side.

Pivot to the right by stepping with the left foot and pivoting off the ball of the right foot. Rotate the hips and shoulders into the direction of the block. The fingers of the left hand should be open and behind the baton.

Place the baton perpendicular to the opponent's striking surface to absorb the impact of the blow.

Bend the elbows slightly to help absorb the impact of the blow. The arms should give with the strike of the blow.

Left Block. Marines execute a left block to deter an attack directed at the head, neck, flank, or hip. The opponent's blow can be delivered by a foot, knee, fist, or elbow. To execute the left block, Marines—

Thrust the baton in a vertical position to the left side.

Pivot to the left by stepping with the right foot and pivoting off the ball of the left foot. Rotate the hips and shoulders into the direction of the block. The fingers of the left hand should be open and behind the baton.

Place the baton perpendicular to the opponent's striking surface to absorb the impact of the blow.

Bend the elbows slightly to help absorb the impact of the blow. The arms should give with the strike of the blow.

Middle Block. Marines execute either a left or right block to deter an attack directed at the face, throat, chest, or abdomen. To execute the middle block, Marines—

Thrust the baton in a vertical position straight out in front of the body. The fingers of the left hand should be open and behind the baton.

Place the baton perpendicular to the opponent's striking surface to absorb the impact of the blow. The baton should be held with the left hand forward of the right.

Bend the elbows slightly to help absorb the impact of the blow. The arms should give with the strike of the blow.

Restraining Technique

The strong-side armlock is used to restrain an opponent who is not compliant. To execute the strong-side armlock, Marines—

Use the right hand to run the baton up under the opponent's left armpit, parallel to the ground.

Use the right foot to step forward at a 45-degree angle to the left side of the opponent. The baton is across his forearm.

Use the right hand to drive the baton forward and up so the action bends the opponent's arm behind his back. At the same time, continue moving around him to get behind him.

Place the baton on the opponent's forearm with the thumb and/or fingers and apply pressure to his forearm. At the same time, grasp the other end of the baton with the left hand.

Pull up on the low end of the baton with the right hand. At the same time, push down on the top end of the baton with the left forearm, reaching around with the left hand to grasp the opponent's biceps or shoulder.

Continue exerting downward pressure with the left forearm while pulling back on the opponent's biceps with the left hand. This places the opponent in a position where he is controlled and can be moved.

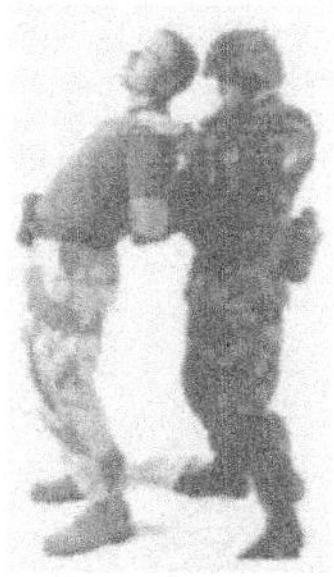

Apply pressure with the foot against the bend in the opponent's leg above his calf. This lowers the opponent to the ground rather than throwing him to the ground and risking severe injury.

Striking Target Areas

Marines must avoid striking an opponent in the head, neck, or other bony parts with the baton because this is considered deadly force and can lead to serious bodily injury or death. Instead, the legs, arms, and buttocks are target areas that are considered nonlethal.

Legs. Primary targets are the thighs and lower legs. Avoid striking the knee and ankle joints because this can cause permanent damage.

Arms. Primary targets are the upper arms. Avoid striking the shoulder and elbow and wrist joints because this can cause permanent damage.

Buttocks. Primary targets are the buttocks. Avoid striking any other part of the torso, including the chest, rib cage, spine, tail bone, and groin because strikes to these areas can cause permanent damage or death.

One-Handed Striking Techniques

One-Handed Forward Strike. A forward strike follows either a horizontal line or a downward diagonal line using a forehand stroke. To execute the one-handed forward strike, Marines-

Stand facing the opponent with the baton carried in a one-handed cany.

Place the right hand palm up, swing the baton from right to left, and make contact with the opponent.

One-Handed Reverse Strike. A reverse strike follows either a horizontal line or a downward diagonal line using a backhand stroke. To execute the one-handed reverse strike, Marines—

Stand facing the opponent with the baton carried in a one-handed carry.

Bend the right arm and cross the arm to the left side of the body. The baton should be close to or over the left shoulder.

Place the right hand palm down, swing the baton from left to right, and make contact with the opponent.

Two-Handed Striking Techniques

Two-Handed Forward Strike. This strike is an effective follow-up to a middle block or left block. To execute the two-handed forward strike, Marines—

Pull back with the left hand while driving the right hand forward toward the opponent. The baton should be horizontal to the ground. Power is generated by stepping forward with the right foot and rotating the right hip and shoulder into the strike.

Contact the opponent with the end of the baton.

Two-Handed Reverse Strike. This strike is an effective follow-up to a middle block or left block. To execute the two-handed reverse strike, Marines—

Pull back with the right hand while driving the left hand forward toward the opponent. The baton should be horizontal to the ground. Power is generated by stepping forward slightly with the left foot and rotating the left hip and shoulder into the strike.

Contact the subject with the end of the baton.

Front Jab. This strike is effective for countering a frontal attack. It can also be executed as a quick poke to keep a subject away. To execute the front jab, Marines thrust both hands forward in a quick jab. The baton is held either horizontal to the ground or at a slight downward angle.

Rear Jab. This strike is effective for countering a bear hug from the rear. To execute the rear jab, Marines thrust both hands rearward in a quick jab. The baton is held either horizontal to the ground or at a slight downward angle.

APPENDIX A

PUGIL STICK TRAINING

A pugil stick is a training device used to simulate a rifle bayonet so that effective, but safe, training can be conducted to build proficiency of rifle bayonet techniques. Pugil stick training builds on the techniques used to throw punches. Pugil stick training is the only "full contact" training provided to Marines in the Close Combat Program. Pugil stick training teaches Marines to function when faced with stress and violence, and it prepares them to deliver a blow and take a blow. It also provides them with the physical and mental skills vital to success on the battlefield.

1. Pugil Stick Training

Design

A pugil stick consists of a stick wrapped in padding at both ends that can be gripped like a rifle. The pugil stick is approximately the same weight and length of an unloaded rifle with a bayonet attached.

Holding the Pugil Stick

The pugil stick is held in the same manner as the sen ice rifle. All movements come from the basic warrior stance. To hold the pugil stick correctly, Marines—

Use the right hand to grasp the lower end of the pugil stick overhanded.
Use the left hand to grasp the upper end of the pugil stick underhanded.
Use the right forearm to lock the lower end of the pugil stick against the hip.
Orient the blade end of the pugil stick toward the opponent.

Safety Equipment

The following safety equipment must be worn during any pugil stick bout.

Groin Protection. Groin protection protects the groin from an accidental blow. It should be pulled high around the waist to protect the groin area, with the concave portion against the body.

Flak Jacket. The flak jacket provides protection to the body. It is worn completely fastened.

Neck Roll. The neck roll prevents whiplash if Marines receive a blow to the head. The neck roll further supports the head and protects the neck from blows. The neck roll is worn above the flak jacket and below the helmet. The tied end of the neck roll faces the front.

Helmet. A regulation football helmet protects the face and head. The helmet must fit snugly and the chin straps must be adjusted and snapped.

Mouthpiece. The mouthpiece is worn on the upper teeth to protect the teeth.

Gloves. Marines wear gloves to protect the hands if pugil sticks do not have gloves built-in.

Pugil Stick Screening

Prior to pugil stick training, instructors must ask participating Marines the following questions. If a Marine answers "yes" to any question, he must see the corpsman or a competent medical authority who evaluates him and determines whether or not the he can participate in the training.

Are you on light duty?
Are you restricted to running shoes by a corpsman or a doctor?
Have you fought in a pugil stick bout within the last 7 days?
Have you received a blow to the head within the last 7 days?
Have you had a concussion within the last 6 months?
Have you had dental surgery within the last 24 hours?
Do you have stitches or staples on your body?
Have you had a shoulder or head injury within the last 5 years?
Are you taking a prescription drug?
Do you have an ear infection or current sinus infection?
Have you had a broken bone within the last 6 months?

Safety Personnel

The following safety personnel are required to conduct pugil stick training:

One close combat instructor must officiate the bout.
One close combat instructor trainer, commissioned officer, or staff noncommissioned officer must be in the training area to serve as range safety officer.
One corpsman must be in the training area.

Training Area

To prevent injury, Marines train on areas with soft footing (i.e., sand or grass). Training mats are not recommended because feet can stick to the mats, prohibiting movement or causing joint injuries by twisting a knee. Bouts should not take place on a hard surface area; e.g., a flight deck or parking lot. A boxing ring may be used to conduct pugil stick bouts; ring dimensions can vary as long as there is ample room to execute the techniques in the training area.

Second Impact Syndrome

Second Impact Syndrome occurs when a second blow to the head produces a second concussion that occurs within 1 week following a previous concussion (before recovery from the first concussion). Second Impact Syndrome causes rapid brain swelling and can cause death. Therefore, there must be 7 days between pugil stick bouts to reduce the risk of severe injury resulting from Second Impact Syndrome. The 7 day separation between pugil stick bouts significantly reduces the possibility of injury, particularly in someone who may have suffered a brain injury or concussion but shows no symptoms.

Any Marine who experiences headaches or the following symptoms after training must be examined by appropriate medical personnel:

Blurred vision.
Ringing in the ears.
Dilation of the pupils.
Slurred speech.
Bleeding from ears or mouth.
Swelling in head or neck area.
Any unnatural discoloration in head or neck area.

The Marine should not be allowed to participate in pugil stick training or any other activity where a heavy blow might be sustained for a minimum of 7 days after the headache or other symptoms have subsided.

Safety Measures

The following safety measures must be followed:

Gear must be worn properly throughout training.
Contact lenses or glasses will not be worn.
False teeth will be removed from the mouth.
Nothing will be worn around the neck except the neck roll.

Competition among groups of Marines is authorized as long as it does not overshadow training objectives or compromise safety procedures.

Safety and proper techniques are paramount.

Safety is more important than competition.

2. General Rules and Regulations Governing Pugil Stick Bouts

Instructors and Support Personnel Requirements

At a minimum, one close combat instructor observes a pugil stick bout. For safety purposes, it is better to have two instructors judging a bout because each instructor can fully observe each of the fighters. The best position for observation is to the right of a fighter. This allows the instructor to see the fighter's facial expression and body movement. The instructor's position must not interfere with the fight.

In addition, one commissioned or staff noncommissioned officer (to serve as range safety officer) and one corpsman will be in the training area.

Prior to the Bout

Prior to the bout, Marines—

Are paired according to height, weight, and gender.

Wear the proper safety equipment.

Once Marines have properly donned the safety equipment, they wait for instructions from the close combat instructor. On the command of the close combat instructor, a pair of Marines will enter the ring. Once in the ring, the close combat instructor inspects each Marine for the proper safety equipment.

Note: There should be enough gear so that when two Marines are training, two other Marines can be donning safety equipment for the next bout.

During the Bout

The bout begins when the close combat instructor blows the whistle. All strikes are directed above the waist. Upon hearing a whistle blast, all fighting immediately ceases.

Stopping a Bout

There are two reasons for stopping a bout: delivery of a scoring blow or an unsafe condition. The close combat instructor trainer, close combat instructor, or the range safety officer may stop the bout at any time an unsafe condition is observed.

Scoring Blow. A scoring blow is an offensive technique delivered to a vulnerable area of an opponent with sufficient force and precision to be considered as a disabling or killing blow. Scoring blows are not judged solely on the degree of force with which a blow is delivered, but on the accuracy and techniques employed. A scoring blow is defined as—

A straight thrust with the blade end of the weapon (red end of the pugil stick) to the opponent's face mask or throat.

A slash to the side of the opponent's helmet (below the ear) or neck with the red end of the pugil stick.

A heavy blow to the opponent's head with an authorized technique (i.e., buttstroke, smash) using the butt (black end) of the pugil stick.

When a scoring blow is delivered, the close combat instructor blows the whistle to stop the bout.

Unsafe Condition. The bout will be stopped as soon as an unsafe condition exists. An unsafe condition exists when a Marine is unable to defend himself, loses his balance and falls down on one or two knees or falls down completely, shows instability (e.g., buckling at the knees), loses muscular tension in his neck and his head snaps back or to one side, or appears disoriented. An unsafe condition also exists if a Marine lets go of one end of the pugil stick, the equipment (e.g., helmet, neck roll) falls off, or a Marine fails to use the proper techniques.

If any of these conditions occur—

The close combat instructor trainer, close combat instructor, range safety officer, or anyone supervising training stops the bout and separates the two Marines.

The corpsman evaluates the possibly injured Marine to determine if the fight can continue. The corpsman—

- Ensures that he is alert and responsive, both verbally and physically.
- Talks to him to see if he is coherent.
- Makes certain he comprehends and replies to verbal questions.
- Makes certain his speech is not slurred.
- Checks his physical signs.
- Ensures his eyes are focused and not dazed or glazed.
- Ensures his legs are not wobbly or shaky.

The pugil stick is not used as a baseball bat. The use of unauthorized techniques will result in expulsion from the bout. Only techniques taught in bayonet training are authorized. These techniques include-

Straight thrust.
Buttstroke (horizontal and vertical).
Smash.
Slash.
Parry.
Blocks (low, high, left, right).

After the Bout

A third whistle blast is used prior to identifying the Marine who delivered the killing blow.

A fourth whistle blast is used prior to demonstrating the killing technique used by the Marine who won the bout.

3. Directions for Making Pugil Sticks

Close combat instructor trainers are taught to make pugil sticks to standard specifications at school. It is recommended that all pugil sticks be made under the supervision of a close combat instructor trainer. Until the Marine Corps acquires a standardized pugil stick that units can order through the supply system, instructors use the following information and instructions to make pugil sticks locally.

Supplies

The following supplies are needed to make one pugil stick:

One 45" by 2" circular, oak, wood dowel.
Four 9" wide by 27" long foam pad pieces to make two inner handguards of double thickness.
Four 11" wide by 19" long foam pad pieces to make two outer handguards of double thickness.
Canvas material to cover inner and outer handguards.
One foam pad cut 1/4" thick by 7" wide by 13" long to make the centerguard.
One 8" wide by 14"long piece of canvas material to make a pugil stick sock.
Canvas material to make two pugil stick end caps. Ideally, two different color fabrics should be used to simulate the bayonet and buttstock ends of a rifle. Common colors for end caps are black and red. Each end cap is a different color to help the officiating certified close combat instructor determine the scoring blow during the pugil stick bout.
Two 1/4" thick by 11" wide by 54" long foam pad pieces to make two ends.
Two foam pads cut 1/4" thick by 3" wide by 14 1/2" long to make two end plug inserts.
One can of 3M Photo MountQ spray adhesive (NSN 8040-01-120-4009)to adhere foam padding to the stick. (Spray adhesive is available from self-service.)
One roll of duct tape or riggers tape (NSN 7510-00-266-5016) to protect foam pad and

canvas from wear and to reinforce gluing action of spray adhesive. (Duct tape or riggers tape is available from self-service.)
One roll para-cord or 5-50 cord to tie both the inner and outer handguards together.
One roll of thread strong enough to sew canvas material together.
Sixteen grommets, one for each of the four corners of the inner and outer handguards.
One foam sleeping mat (NSN 8465-01-109-3369). Use unserviceable foam sleeping mats obtained from Defense Reutilization and Marketing Office.

Tools

The following tools are needed to make a pugil stick:

Tape measure, ruler, or yardstick.
Scissors.
Grommet machine.
Knife.
Sewing machine.

Directions

To make a pugil stick, Marines perform the following steps.

Preparing Materials

Step 1. Cut foam sleeping mat into the dimensions listed above. Cut mats to double thickness.

Step 2. Use spray adhesive to adhere the two 9" by 27" pieces of foam together. Use the spray adhesive to adhere the two 11" by 19" pieces of foam together. Let adhesive dry before using the double-thick foam for handguards.

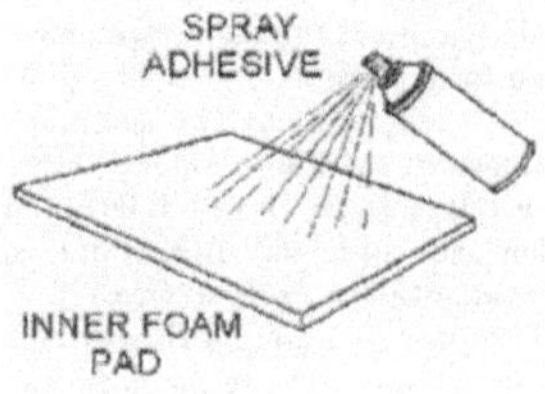

Step 3. Cut canvas material to make two coverings for the inner handguards. Cut material to make two coverings for the outer handguards. Two inner handguards and two outer handguards are required for each pugil stick.

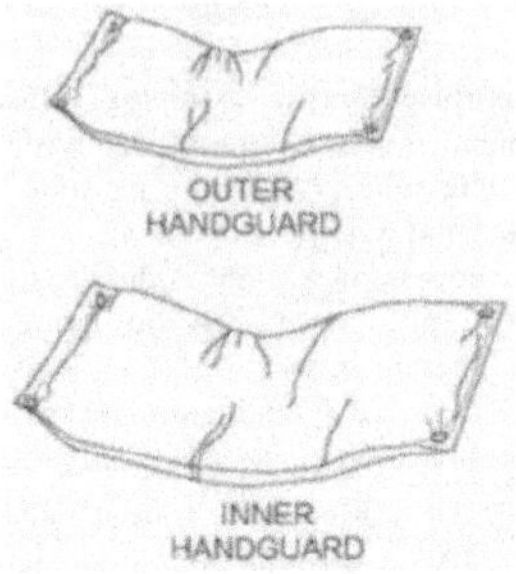

Step 4. Use a sewing machine to make a sleeve out of the 9" by 27" canvas. Sew down three sides. Insert foam pad inside the canvas sleeve and then sew down the fourth side. Use the grommet machine to create holes reinforced with metal grommets in each of the four corners. Once finished, the inner handguards arc complete.

Note: It is recommended that the Fabric Repair Shop be used to sew canvas sleeves and apply grommets.

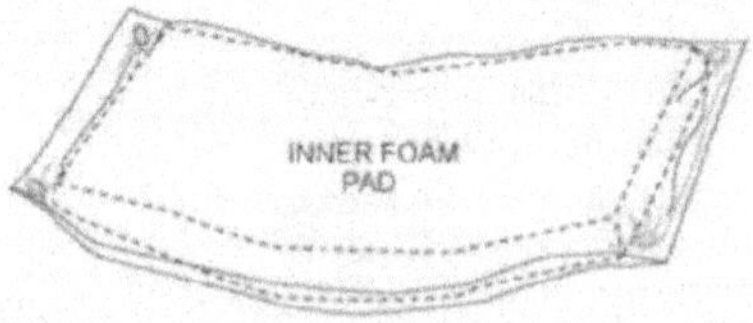

Step 5. Repeat step 4 to make the other inner handguard and both outer handguards.

Constructing Ends, End Plugs, and End Caps

Step 1. Apply spray adhesive on approximately 8 inches of one end of the dowel. This adhesive serves to glue the foam pad to the wooden dowel.

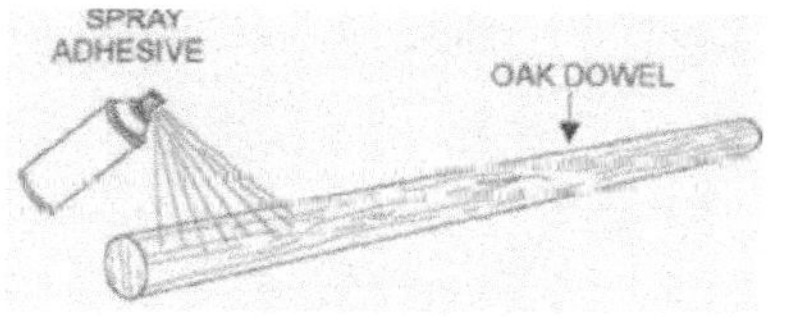

Step 2. Apply spray adhesive on one side of the 11" by 54" foam pad. This causes the foam pad to stick to itself when it is rolled around the end of the wooden dowel. Ensure 3 inches of the rolled foam pad extends past the wooden dowel for stability.

Step 3. Apply spray adhesive on one side of 3" by 14 1/2" foam pad. Roll pad into a tight cylinder.

Step 4. Apply spray adhesive on the exposed end of the dowel and fill the hole in the end with the end plug. The end plug prevents the wooden dowel from protruding beyond the edge of the rolled foam padding. This is an important safety precaution to prevent injury.

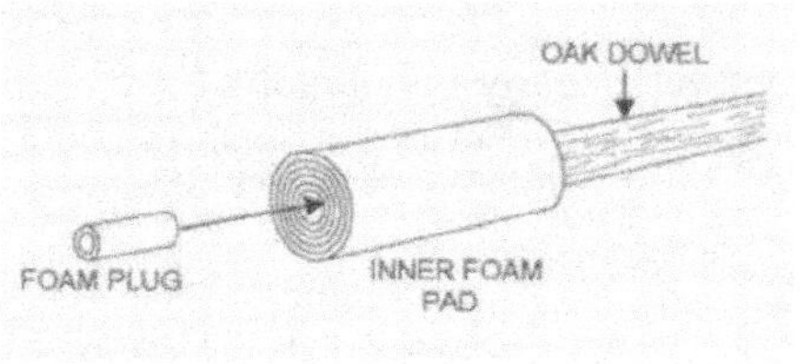

Step 5. Use duct tape or riggers tape to secure and reinforce the foam pad once the end plug is inserted in the end.

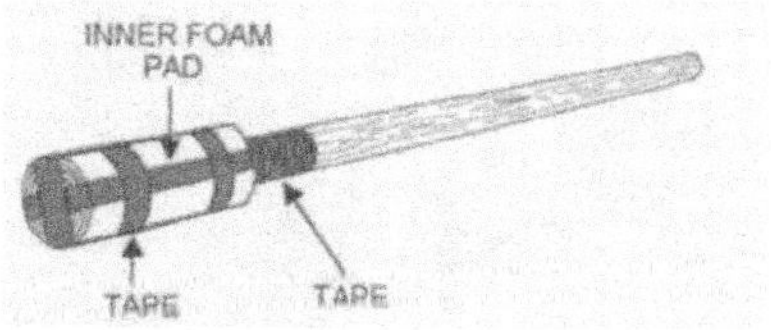

Step 6. Measure the dimensions of the ends (typically, the length of the cylinder should be 12 inches long to allow for taping). Cut and sew canvas material in the shape of a cylinder with one open end. Once complete, this becomes the end cap. The end cap is a colored sleeve that covers the rolled and taped foam pad at each end of the pugil stick.

Note: It is recommended that the Fabric Repair Shop be used to make the end cap.

Step 7. Slide one of the end caps over the rolled foam pad and tape it to the dowel.

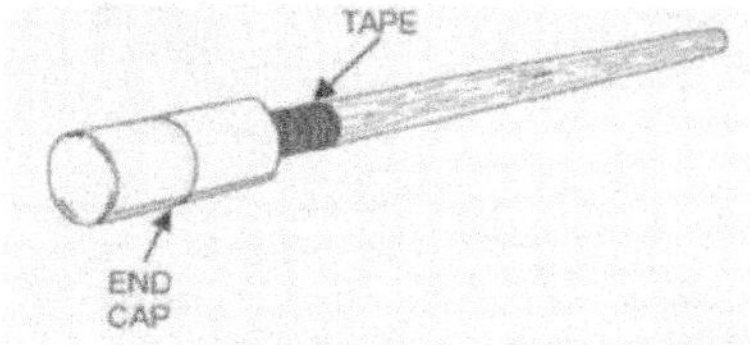

Building the Centerguard

Step 1. Spray one side of the 7" by 13" foam pad with spray adhesive and wrap it tightly around the center of the dowel.

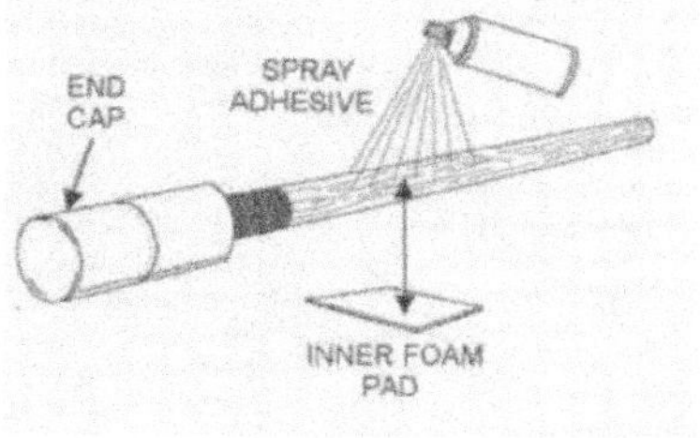

Step 2. Wrap duct tape or riggers tape around the foam pad to secure.

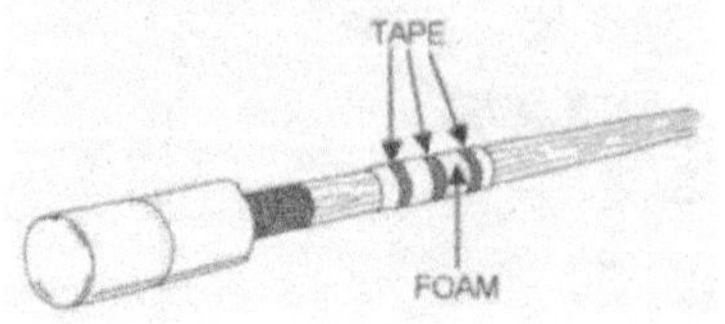

Step 3. Use a sewing machine to make a tubular pugil stick sock out of the 8" by 14" piece of canvas. Slide the pugil stick sock over the rolled foam pad and secure it with the duct tape or riggers tape. Duct tape or riggers tape protects the edges of the centerguard and reinforces the gluing effects of the spray adhesive.

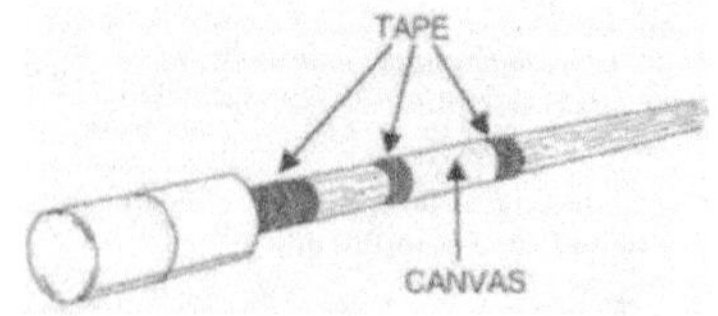

Note: It is recommended that the Fabric Repair Shop be used to make the pugil stick sock.

Finishing the Opposite End Cap

After one end cap is complete and the centerguard is finished, then repeat steps 1 through 7 in Constructing Ends, End Plugs, and End Caps for the other end of the pugil stick. Ensure that the opposite color end cap is used.

Fastening Handguards

Step 1. Place the inner handguards (longer piece) around the dowel. The inner handguard is placed on first because it protects the wrist. One inner handguard is needed for the left hand and one for the right hand.

Step 2. Fold the inner handguard around the pugil stick and run the para-cord or 5-50 cord through the metal grommets.

Step 3. Tighten down the corner ends of the inner handguards by pulling on the para-cord or 5-50 cord.

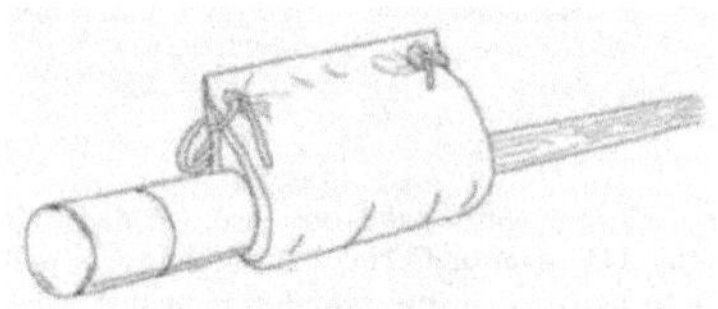

Step 4. Tie off the cord with a square knot above the metal grommets.

Step 5. Place the outer handguards (shorter piece) over the inner handguards (longer piece). Repeat steps 1-4 for the other handguard. The outer handguards provide additional protection to the knuckles and fingers and provides additional stability for the inner handguards.

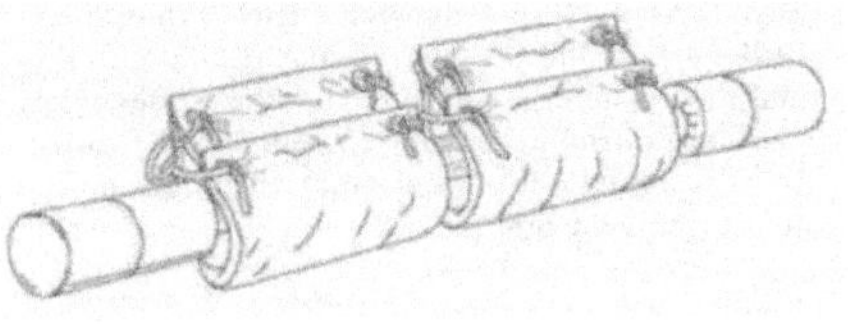

Serviceability Inspection

Prior to any training, all pugil sticks are inspected to ensure they are fit for use. If any of the following conditions exist, the pugil stick is unsafe and will not be used:

Stick protrudes through the end caps.
Frayed or worn edges exist on canvas surfaces.
The stick can be felt through the pads or the padding is too soft.
End caps or handguards are not securely attached.

(reverse blank)

APPENDIX B

SAFETY PRECAUTIONS DURING TRAINING

1. General Safety Precautions

When training close combat techniques, certain safety precautions must be adhered to to prevent injuries.

Most training should be conducted on a training area with soft footing such as a sandy or grassy area. If training mats are available, they should be used. A hard surface area is not appropriate for close combat training.

All techniques should be executed slowly at first. Marines can increase the speed of execution as they become more proficient. Marine-on-Marine training that requires contact (chokes, throws, ground fighting, and unarmed restraints and manipulation) should not be executed at full force or full speed.

If a technique is applied to the point that a Marine is uncomfortable, the Marine must "tap out." This indicates immediate release of the pressure being applied or to immediately stop the technique. The Marine "taps out" by firmly tapping his hand several times on any part of the opponent's body that will get his attention or by saying *stop.*

Second Impact Syndrome occurs when a second concussion develops within hours, days, or weeks following a prior concussion (and before recovery from the first concussion). Second Impact Syndrome causes rapid brain swelling and can cause death. Marines who experience headaches or other symptoms following training must be examined by appropriate medical personnel. These symptoms can include, but are not limited to, blurred vision, ringing in the ears, dialation of the pupils, bleeding from the ears or mouth, slurred speech, swelling in head or neck area, or any unnatural discoloration of head or neck. They should not be allowed to participate in pugil stick training or any other activity where a heavy blow might be sustained for a minimum of 7 days after the symptoms have subsided.

2. Safety Precautions for Individual Techniques

Falls

When training Marines to fall, they should progress from the ground, to a kneeling or squatting position, and then to a standing position. This ensures they are comfortable and understand the technique before progressing to executing falls from a higher profile. This instructional technique vastly reduces the risk of injury.

Strikes and Punches

When training Marines to strike and punch, they begin by executing the techniques "in the air." As they become more proficient, they execute strikes on equipment (when available) such as an air shield or a heavy bag. At no time should they be permitted to execute strikes on another student.

Chokes and Ground Fighting

Chokes

When training Marines to execute chokes, they will not apply pressure to the opponent's throat during training because the trachea and windpipe can be crushed. During training, Marines should practice the proper procedures for blood chokes. They should never execute air chokes. At no time during training should a choke be applied at full force or full speed. No choke should be held for longer than 5 seconds.

Unarmed Restraints and Manipulation

When training Marines to execute unarmed restraints and manipulation techniques, they utilize slow and steady pressure. Never apply these techniques at full force or at full speed.

Pugil Stick Training

A blow to the head during training will bruise the brain. A second blow to the head can cause death. This is known as the Second Impact Syndrome. There must be 7 days between pugil stick training to prevent injury or death.

CONTENTS

INTRODUCTION

0001. PURPOSE. To promulgate policy and implementing instructions, concerning physical fitness and body composition in the Marine Corps.

0002. STATUS

1. The policies and procedures in this Manual apply to HQMC staff agencies and COs who are authorized to issue directives concerning this program.

2. The CMC (C 472TP) must authorize any deviation from the instructions in this Manual.

0003. RESPONSIBILITY. The CMC (C 472TP) is responsible for the accuracy, currency, modification, and distribution of this Manual. HQMC staffs and COs are responsible for the timely entry of changes and the physical maintenance of copies of this Manual.

0004. ALLOWANCES

1. Submit requests for changes in allowances of this Manual to the CMC (C 472TP). Submit requests per reference (b).

2. Request missing pages by requisitioning the basic Manual and pertinent changes per reference (b).

0005. ORGANIZATION

1. This Manual is organized into chapters identified by an Arabic numeral as listed in the overall contents.

2. Paragraph numbering is based on four digits. The first digit indicates the chapter; the next digit, the section; the final two digits the general major paragraph number; and the combinations which follow the decimal point, the subparagraph number; e.g., 3103.3a(2) refers to chapter 3, section 1, general major paragraph number 03, subparagraph 3a(2).

3. Pages are numbered in separate series by chapter number, with the chapter number preceding each page number; e.g., the fourth page of chapter 2 is shown as 2-4.

4. A glossary of terms is provided in appendix A to assist users with definitions of the numerous terms and acronyms throughout this Manual.

0006. CHANGES. Printed changes are posted per reference (c). Interfile changes in the basic Manual in consecutive order and complete the Record of Changes page.

CHAPTER 1

PHYSICAL CONDITIONING PROGRAM (PCP)

CHAPTER 1

PHYSICAL CONDITIONING PROGRAM (PCP)

1100. SCOPE. Every Marine must be physically fit, regardless of age, grade, or duty assignment. Fitness is essential to the day-to-day effectiveness and combat readiness of the Marine Corps. Furthermore, physical fitness is an indispensable aspect of leadership. The habits of self-discipline required to gain and maintain a high level of physical fitness are inherent to the Marine Corps way of life and must be a part of the character of every Marine. Marines who are not physically fit can be a detriment to the readiness and combat efficiency of their unit. Accordingly, every Marine will engage in an effective PCP on a continuing and progressive basis.

1101. RESPONSIBILITY. In order to maintain continuity of standards, the procedures in this Manual will be used for developing and administering the COs conditioning, testing, and remedial programs, and to clarify Body Composition Program (BCP) policies (formerly known as the Weight Control Program). It is the responsibility of COs to provide a fair evaluation of Marines' physical fitness and body composition through the procedures outlined in this Manual. When necessary, an Appropriately Credentialed Health Care Provider (ACHCP), which will be referred to throughout this Manual as a Medical Officer, or "MO", will be required to conduct a thorough physical evaluation; an ACHCP is defined in the Glossary, appendix A of this Manual. This evaluation will include at a minimum, a Body Composition Evaluation and a determination of whether or not the weight and body fat gain is due to an underlying cause or associated disease.

1102. OBJECTIVES

1. Overall Program Focus. The Unit's PCP should focus on combat conditioning, health, fitness, and unit cohesion rather than preparation exclusively for the semi-annual Physical Fitness Test (PFT). The Program should reflect the following objectives:

a. To contribute to the health and well-being of every Marine through regular exercise, fitness testing, and health education.

b. To develop Marines who are physically capable of performing their duties in garrison and in combat.

c. To develop in Marines a reserve level of physical fitness and endurance that will enhance their chances of success in a combat situation.

d. To provide a medium for developing the individual Marine's self-confidence and desire to excel, thereby enhancing the unit's overall discipline, morale, esprit de corps, and combat readiness.

2. CO's Focus. COs should strive to achieve the primary objective of combat readiness by emphasizing combat conditioning at the unit level. The CO should ensure this objective is reflected in the training plans and schedules, affording the individual Marine an opportunity to achieve a competent level of combat fitness.

3. Command Physical Training (PT) Representative's Focus. The CO will assign in writing a Marine SNCO (normally from the Operations and Training Section), as the Command PT Representative. This Marine will be the commander's senior representative in the administration of Marine Corps policy related to physical conditioning, testing and remedial programs. Of particular emphasis, this Marine will be responsible for:

a. Conduct of the PFT. Responsible to the CO for selection and training of PFT monitors and for all matters related to the PFT (safety, course selection, recording, etc.).

b. Body Composition Evaluations. To ensure accuracy in body composition evaluations, the **only** personnel authorized to conduct these measurements are the Command PT Representative and his/her designate.

1103. PHYSICAL CONDITIONING. The definition of physical fitness is subjective to those defining it. The Marine Corps considers physical fitness to be the ability of a Marine to meet the physical demands of any combat or duty situation without undue fatigue. To achieve this state of physical fitness, physical conditioning should incorporate the components and principles listed below.

1. Components of Physical Conditioning. To capitalize on those components that can benefit conditioning efforts, the following categories of exercises should be included in both individual and unit PCPs:

a. Strength. Muscular strength refers to the ability of the muscular system to move the body through resistance. Many associate strength training with progressive resistive exercises using weights and machines. However, the ability of a Marine to effectively handle their own body weight should be a prerequisite before integrating strength training with machines into their program. This can be accomplished through the Daily 16 Program described in paragraph 1103.3a, and applying the principles listed in paragraph 1103.2. Strength training can be broadly separated into two categories, general and specific.

(1) General Strength Training. This type of training strengthens the muscular system by focusing on a full body workout for strength and size. In this type of training, the major muscle groups are exercised without a specific task or functional goal in mind. This type of strength training contributes to overall health.

(2) Specific Strength Training. This type of strength training is task specific. For example, Marines desiring to climb a rope better would do rope climb training wearing body armor, and focus their strength training on muscles involved in rope climbing. A company of Marines expecting to operate in hilly terrain would focus their strength training primarily on lower body strength.

b. Endurance. Two types of endurance conditioning are needed for a Marine to meet the physical demands of combat, aerobic and anaerobic.

(1) Aerobic Endurance. Aerobic activity, meaning "in the presence of oxygen," is categorized by physical demands that are sub-maximal (not an "all out" effort) and involve activity that is continuous in nature (lasting more than 3-5 minutes). Two examples are road marching and long distance running.

(2) Anaerobic Endurance. Anaerobic activity, meaning "without oxygen," is categorized by physical demands that are high intensity and of shorter (less than 2-3 minutes) duration. Examples are rope climbing, most forms of weight lifting, and running short, quick distances.

c. Mobility. Mobility conditioning is geared towards improving quality of movement. Quality of movement depends on the following:

(1) Posture
(2) Balance and stability
(3) Agility
(4) Coordination
(5) Power
(6) Speed
(7) Flexibility

2. Principles of Physical Conditioning. There are several different principles to consider when developing an effective PCP:

a. Progression. Conditioning programs must incorporate a systematic means to increase training load.

b. Regularity. To realize a conditioning effect, training programs must conduct physical conditioning sessions at least 3-5 times per week.

c. Overload. Only when the various systems of the body are overloaded will they become able to handle greater load.

d. Variety. Varying a program from time to time maintains interest and prevents staleness.

e. Recovery. Essential for allowing the systems overloaded during conditioning to adapt and become stronger.

f. Balance. Balanced conditioning programs ensure all the components of physical fitness conditioning (strength, endurance, and mobility) are properly addressed.

g. Specificity. Conditioning that is specific in nature yields specific gains. For example, stationary bike riding is of little value in improving running.

3. Specific PCPs. Units that strive to augment their PCP should use innovative combinations of the types of exercise defined below:

a. Daily 16 Program. The Daily 16 Program is a comprehensive series of warm-up, conditioning, and cool-down exercises replacing the former Daily 7 Program. This all-encompassing program can be incorporated into any unit aerobic or anaerobic conditioning session, or can be used as a conditioning session in of itself. Appendix B contains the Daily 16 Program in detail.

(1) Warmup Exercises. The Daily 16 warm-up exercises facilitate gradual distribution of blood flow to the muscles, preparing both the cardiovascular and musculoskeletal systems for the exercise session, by effectively targeting both the upper and lower body. The increased blood flow to the muscles produces a warming effect, increasing the elasticity of the muscles and connective tissue, which is believed to reduce injury.

(2) Conditioning Exercises. The Daily 16 conditioning exercises provide a total body workout through the proper execution of traditional calisthenics. Conditioning exercises can be used as a workout session in and of itself, or to augment the main fitness event (e.g., squad ability run or obstacle course).

(3) Cool-down Exercises. The Daily 16 cool-down exercises (which are basically the same as the warm-up exercises) allow the body to gradually return to the pre-exercise state.

b. Occupational Conditioning. This conditioning is comprised of general physical conditioning exercises that will develop and maintain strength, endurance, and the physical skills necessary to sustain a Marine during combat. Good examples of occupational conditioning that prepares Marines to successfully handle the demands of their particular MOS/billet are Physical Readiness Training (PRT), progressive load-bearing marches, martial arts training, dry net training, Military Operation in Urban Terrain (MOUT) training, combat water survival training, obstacle course, and confidence course.

c. Competitive Conditioning. Competitive conditioning activities consist of teams or individuals competing against an opponent to win. This includes a combination of sports and military skills designed to foster the unit's combat readiness, competitive spirit, and esprit de corps. Some good examples are speed-marches reaction courses, orienteering, water-can re-supply relays, and stretcher or fireman carry relays.

d. Remedial Physical Conditioning Program (RPCP). All Marines who fail the PFT, demonstrate poor physical performance in MOS-related skills, or who are assigned to the BCP will be assigned to a RPCP. Remedial physical conditioning is a process by which Marines are conditioned in a progressive manner to meet prescribed standards. The goal is to provide challenging conditioning sessions, using the spectrum of aerobic conditioning, resistance conditioning, and other related exercises that will afford the Marines an opportunity to rebuild themselves after a weight and body fat increase, injury/illness, pregnancy, or a period that lacked a structured fitness program. The philosophy of remedial conditioning is not punitive in nature, but is intended to encourage improvement in overall fitness or to supplement a BCP. Comprehensive guidance for the development of a command directed RPCP is contained in paragraph 2400 and appendix C of this Manual.

e. Alternate Aerobic Conditioning. In cases of lower limb injury or related medical conditions that prevent running or hiking, Marines may opt to perform other low impact activities to supplement the requirement for cardiovascular conditioning. Swimming, deep water running, cycling, cross-country skiing, stair stepping, and rowing are excellent examples of low impact, endurance exercises that provide good augmentation or substitution for running or hiking regimens. For example, the advantages of swimming include the use of all major muscle groups during exercise, reducing lower body stress due to partial body weight being supported by the water, and the absence of impact on bones and joints. Additionally, the body's position in the water increases the efficiency of the circulation back to the heart during exercise. Although these types of aerobic conditioning alternatives may not improve overall running ability (e.g., speed and endurance), it can enhance a current running or hiking regimen, while minimizing related injuries.

1104. MARINE CORPS PCP REQUIREMENTS

1. Individual Requirements. The PCP requires every Marine to:

a. Participate in a minimum of 3 hours of physical fitness conditioning per week; 5 hours is encouraged if the unit's operation tempo can accommodate such a program.

b. Be tested semi-annually - PFT. (Reserves tested annually.)

c. Obtain a minimum 3rd Class score on all semi-annual PFTs.

1105. <u>CO's ACTIONS</u>. The PCP must provide opportunities for development through supervised individual or group effort, and sustained by common sense leadership at all unit levels. In order to develop and maintain the desired level of fitness, COs will:

1. Schedule a minimum of 3 hours per week of mandatory individual or group physical conditioning sessions. COs are encouraged to use normal working hours to satisfy this training requirement. It is recommended that each conditioning session be a minimum of 60 minutes in duration.

2. Establish, implement, and maintain the PCP requirements as stated in this Manual.

3. Ensure all Marines participate in exercises or activities commensurate with their medical qualifications and limitations.

4. Ensure proper administration and tracking for all Marines in a light and limited duty status (e.g., Morning Reports).

5. Ensure the unit's PCP is not constructed and implemented solely to prepare for a PFT, but that it emphasizes physical conditioning for combat. This distinction should be reflected in the unit training schedules.

6. Conduct a semi-annual PFT for all Marines per the instructions contained in chapter 2 of this Manual.

7. Place and track all Marines who fail the PFT on a RPCP per the guidance set forth in paragraph 2400 and appendix C.

8. Immediately process Marines for a BCP assignment when identified as not in adherence with weight and body fat standards and do not meet Physical Performance Evaluation criteria.

9. Place and track all Marines assigned to a BCP on a RPCP for the duration of their 6-month BCP assignment.

10. Employ Semper Fit resources to assist in the execution of the PCP, RPCP, and BCP. Semper Fit Fitness Centers staffed by qualified professionals, and the Semper Fit website (www.usmc-mccs.org) provide expert guidance in the areas listed below. The fitness centers, available at most major Marine Corps installations, can provide the following services:

    a. Fitness assessments and exercise prescriptions*;

    b. Exercise orientation;

    c. Group exercise opportunities;

    d. Instructional skill development; and

    e. Health promotion programs.

*Note: Personal fitness assessments conducted by Semper Fit personnel (e.g., body fat testing), while valuable, will not be considered official for BCP testing.

11. Ensure timely and accurate reporting of PFT results and BCP assignments on fitness reports per reference (d), and that correct Unit Diary and SRB entries are made, as appropriate, per references (e) and (f).

12. Ensure the Command retains all completed PFT/BCP files for one year.

CHAPTER 2

PHYSICAL FITNESS TEST (PFT)

SECTION 4: REMEDIAL PROCEDURES

TABLES

CHAPTER 2

PHYSICAL FITNESS TEST (PFT)

2000. GENERAL INFORMATION

1. Purpose. While units are responsible for testing their Marines for combat fitness and readiness as it pertains to their unit's mission, the PFT provides an instrument that measures the collective level of physical fitness Marine Corps wide. It is a measurement of general fitness vice combat readiness and unit/MOS capability. The PFT consists of three events: male Marines will perform dead-hang pull-ups, abdominal crunches, and a 3.0 mile run; and female Marines will complete the flexed-arm hang, abdominal crunches, and a 3.0 mile run. These events are designed to test the strength and stamina of the upper body, midsection, and lower body, as well as the efficiency of the cardiovascular system.

2. Personnel to be Tested. Units will administer the PFT semi-annually (calendar year) for score to all active duty Marines, regardless of age, gender, grade, or duty assignment.

a. PFT Not Required. The condition when a Marine will not be required to conduct a semi-annual PFT:

(1) Personnel on duty in a combat zone; the semi-annual PFT may be waived until their return from the combat zone.

(2) Pregnant Marines medically waived from taking a PFT.

(3) Marines in a light duty or limited duty status, restricting them from participating in the PFT or Partial PFT (PPFT). (Appropriate documentation by a Medical Officer (MO), that assigns light duty or directs a Medical Evaluation Board (MEB) action, is mandatory.)

(4) Marines awaiting separation based on a Physical Evaluation Board (PEB). Reference paragraph 2302 for further details.

b. PFT Requirement for End of Active Service (EAS)/Retirement

(1) Marines are required to take the PFT during the semi-annual period preceding their EAS or retirement date, unless directed by their CO. This date is not the date that terminal leave begins.

For example, a Marine whose EAS or retirement date is September, but is commencing terminal leave in June, will be required to take the semi-annual PFT for the January-June period. This Marine is not required to take the semi-annual PFT for the July-December period.

(2) Retirement/EAS Final Physical Examination. Completion of the required final physical examination, regardless of when it is completed, does not exempt a Marine from taking the PFT. A Marine, who elects to complete their final physical examination 7-12 months prior to EAS or retirement, will still be required to take the semi-annual PFT for that period.

For example, a Marine who completes the final physical in March, but does not EAS or retire until October, will still be required to take the semi-annual PFT for the January-June period.

c. Selected Marine Corps Reserve (SMCR). Reserve Marines are required to conduct an annual PFT for score. Reserve Marines on Active Duty Special Work (ADSW) orders for 30 days or more, can conduct their annual PFT with that unit. All scores are good for 2 years (for cutting score purposes only), if training does not permit testing.

CHAPTER 2

PHYSICAL FITNESS TEST (PFT)

SECTION 1: TESTING PROCEDURES

2100. CONDUCT OF THE PFT

1. Uniform. The authorized uniform for the semi-annual PFT is the USMC established green-on-green T-shirt and shorts, socks, and running shoes. The green-on-green sweat suit may be used in winter climates. Gloves may be worn. Substitutions of the aforementioned uniform are at the discretion of the CO.

2. Sequence of Events. The sequence of PFT events will be left to the discretion of the CO. All PFT events will be conducted in a single session, not to exceed 2 hours in duration. Movement of Marines from one event to the next should allow adequate time to recover, stretch, and drink water.

3. Restriction of Events. The command will not impose restrictions on the performance of any event other than those specifically stated in this Manual. For example, a command will not mandate that Marines must use the overhand grip when executing pull-ups or flexed-arm hang.

4. Monitoring. The Command Physical Training (PT) Representative will monitor the PFT and maintain the PFT and Body Composition Worksheets. Marines will be kept advised of their progress as they are performing each event. The Command PT Representative or designate will call out each repetition for the pull-ups, and time-passed (in intervals) for the flexed-arm hang and abdominal crunch. The Command PT Representative (and additional monitors) will be posted along the 3.0 mile run course. Marines will receive their split time at the halfway mark and finishing time.

5. Logistics. Minimal special-type equipment is required to conduct a PFT.

a. Pull-Up Bar Requirements

(1) Diameter of the bar may range between 1-1¾ inches; athletic tape on the bar is authorized.

(2) The bar must be high enough to allow the tallest Marine's legs to hang straight without touching the ground, when arms are fully extended.

(3) Bench or stool may be used for shorter Marines to initially grasp the bar.

b. 3.0 mile Run Course. The run course must be measured for accuracy and set over reasonably level ground. Course selection criteria must include consideration of the following:

(1) Safety. Operational Risk Management (ORM) measures must be taken to ensure the course does not expose participants to undue risk.

(2) Performance. When possible, the course should be an "out and back" or wide loop course. The run course should not include numerous sharp turns, etc. that would force a participant to slow down excessively to remain on the course.

(3) Location. An indoor or outdoor track is permissible for the conduct of the run event. A determination as to whether the track is a yard or meter track must be made and the track measured to ensure accurate distance, prior to conducting the PFT. Run courses should not require a Marine to do more than 12 laps to cover 3.0 miles, (e.g., a course should not be shorter than 440-yard per lap). Running this event on a treadmill is not authorized.

c. Support Equipment & Personnel (Monitors)

(1) A timepiece (digital or stopwatch) that accurately measures time to the second is required for the flexed-arm hang, abdominal crunch, and 3.0 mile run.

(2) Safety equipment and personnel will be required as prescribed per local SOP.

(3) Adequate supervisory personnel to monitor each event and act as timekeepers.

d. PFT and Body Composition Worksheets. Appendix F contains the worksheet for recording PFT and body composition data. The following information will be recorded on the worksheet and be maintained by the Command PT Representative:

(1) The Marine's name, SSN, rank, age, gender, height, weight, percent body fat (if required), age 46 and Over Risk Factor Screening (if required), Physical Performance Evaluation qualifier (if required), event time(s)/total number of repetitions, event score, total score, and age-based PFT classification.

(2) Points for each event will be awarded according to the PFT Scoring Table located in appendix G.

(3) Minimum acceptable performance requirements by event (time and repetitions), and age group are listed in table 2-1 in this chapter.

(4) Minimum PFT classification scores are listed in table 2-2 in this chapter.

2101. EXECUTION OF THE PFT

1. Individual performance on each PFT event will be executed as follows:

a. Pull-up. The goal of the pull-up event is for the Marine to execute as many accurate and complete pull-ups before dropping off the bar. The procedures are:

(1) This is not a timed event.

(2) Sweatshirts will be removed during the conduct of the pull-up event in order to observe the lockout of the elbows with each repetition.

(3) Assistance to the bar with a step up, being lifted up, or jumping up is authorized. Any assistance up to the bar will not be used to continue into the first pull-up.

(4) The bar must be grasped with both palms facing either forward or to the rear.

(5) The correct starting position begins when the Marine's arms are fully extended beneath the bar, feet are free from touching the ground or any bar mounting assist, and the body is motionless.

(6) The Marine's legs may be positioned in a straight or bent position, but may not be raised above the waist.

(7) One repetition consists of raising the body with the arms until the chin is above the bar, and then lowering the body until the arms are fully extended; repeat the exercise. At no time during the execution of this event can a Marine rest his chin on the bar.

(8) The intent is to execute a vertical "dead hang" pull-up. A certain amount of inherent body movement will occur as the pull-up is executed. However, the intent is to avoid a pendulum-like motion that enhances the ability to execute the pull-up. Whipping, kicking, kipping of the body or legs, or any leg movement used to assist in the vertical progression of the pull-up is not authorized. If observed, the repetition will not count for score.

(9) A repetition will be counted when an accurate and complete pull-up is performed.

b. Flexed-Arm Hang. The goal of the flexed-arm hang event is for a Marine to hang (maintain elbow flexion) for as long as possible. The procedures are:

(1) This is a timed event.

(2) Sweatshirts will be removed during the conduct of the flexed-arm hang event in order to observe when the Marine has completely locked-out her elbows.

(3) Assistance to the bar with a step up, being lifted up, or jumping up to the start position is authorized.

(4) The bar must be grasped with both palms facing either forward or to the rear.

(5) The correct starting position begins when the Marine's arms are flexed at the elbow, the chin is held above the bar and not touching it, and the body is motionless. At no time during the execution of this event can a Marine rest her chin on the bar.

(6) Marines are authorized to drop down below the bar, however, some degree of elbow flexion must be maintained with both arms. Once a Marine's arms are fully extended or the Marine drops off the bar, the clock will stop.

c. Abdominal Crunch. The goal of the abdominal crunch event is for a Marine to execute as many proper and complete crunches within the prescribed time limit. The procedures are:

(1) 2-minute time limit.

(2) On a flat surface, Marines will lie flat on their back with shoulder blades touching the deck, knees will be bent, and both feet will be flat on the deck.

(3) The arms will be folded across the chest or rib cage with no gap existing between the arms and chest/rib cage. Both arms must remain in constant contact with chest/rib cage throughout the exercise. A single repetition consists of raising the upper body from the starting position until both forearms or elbows simultaneously touch the thighs, and then returning to the starting position with the shoulder blades touching the deck.

(4) The buttocks will remain in constant contact with the deck throughout the event. No arching of the lower back or lifting the buttocks is permitted.

(5) An assistant may be used to hold a Marine's legs or feet, at or below the knees in whatever manner that is most comfortable for the Marine. Kneeling or sitting on the Marine's feet is permitted.

(6) A repetition will be counted when an accurate and complete abdominal crunch is performed.

d. 3.0 Mile Run. The goal is for a Marine to complete the measured course as quickly as possible. The procedures are:

(1) This is a timed event.

(2) On the command to start, the two Marines monitoring the event will start their watches simultaneously when the last Marine passes the starting point. The Command PT Representative will remain at the start/finish and the other monitor will take the safety vehicle (with communication capabilities) to the halfway point. Monitors will call out the split or finishing time as appropriate, as each Marine passes.

2. The PFT scoring table is contained in appendix G.

CHAPTER 2

PHYSICAL FITNESS TEST (PFT)

SECTION 2: STANDARDS

2200. PFT STANDARDS

1. PFT Performance Requirements. To successfully pass the PFT, Marines must complete the minimum acceptable performance requirements in each event and achieve an overall combined score for their age group as shown in table 2-1. Marines must be made aware that the minimum performance (points) in each event alone, will not total the points required for a passing score. Additional points must be earned in at least one event in order to achieve a 3rd Class PFT or better, per age group. Failure to meet the minimum requirements in any event constitutes a failure of the entire test, regardless of the total number of points earned for all three events. Marines on light or limited duty will complete the two events that they are medically qualified to participate in, and will receive credit for a PPFT. Paragraph 2202 further details the PPFT requirements.

AGE	PULL-UPS/ FLEXED-ARM	ABDOMINAL CRUNCHES	3.0 MILE RUN (MIN)	TOTAL POINTS	MIN SCORE	ADDNTL' POINTS NEEDED
17-26	3/15(SEC)	50	28(m) 31(f)	105	135	30
27-39	3/15	45	29(m) 32(f)	94	110	16
40-45	3/15	45	30(m) 33(f)	88	88	0
46+	3/15	40	33(m) 36(f)	65	65	0

Table 2-1.--Minimum Acceptable Performance Requirements for PFT/PPFT Events.

2. Individual event scores for both males and females can be viewed in appendix G.

3. PFT Classification Scores. Table 2-2 shows the minimum score required, per age group, to earn each PFT classification score. Marines should be encouraged to continually strive to perform their best and not merely accept minimum performance.

PFT CLASS	AGE GROUPS 17-26	27-39	40-45	46+
1st	225	200	175	150
2nd	175	150	125	100
3rd	135	110	88	65

Table 2-2.--Minimum PFT Classification Scores.

4. PFT scores will be entered into the MCTFS and on FitRpts per paragraph 2203 in this chapter.

2201. ALTITUDE 3.0 MILE RUN. Marines permanently assigned or on TAD to locations at or above 4500 feet above sea level will have adjusted run times. The altitude 3.0 mile run point values are contained in appendix H. Commands are directed to provide newly joined Marines a 30-day acclimation period prior to conducting a semi-annual PFT. Marines scheduled to report to a command at altitude in June or December will complete their PFTs prior to reporting.

> CH 1 2202. PARTIAL PHYSICAL FITNESS TEST (PPFT). Marines who are medically excused from performing a particular event of the PFT as certified by a MO (e.g., light or TLD), are required to perform a PPFT. The PPFT for score requires that Marines complete two of the three PFT events. For the event not conducted, the event score from the previous semi-annual PFT will be used. Event times/repetitions will be awarded points the same way the PFT awards points; reference appendix G to view the scoring table. As with a PFT, failure to meet the minimum requirements in either event constitutes a test failure. Reference table 2-1 to view the minimum acceptable performance requirements for each event.

1. Reporting

a. The PFT and Body Composition Worksheet will indicate the event not taken and whether a light duty, TLD, or PLD status was the reason for the PPFT.

>CH 1 b. Report PPFT performance (pass/fail) for Sgt and above in accordance with reference (e).

c. LCpls and Cpls requiring a composite score will use the score (points) received from the last semi-annual PFT for the event missed.

d. PPFT scores will be entered on the Unit Diary and FitRpt per paragraph 2203.

2. Post-Light or Limited Duty PFT. Marines have the option to either keep the PPFT score taken for the current semi-annual period or to conduct a PFT (e.g., for promotion purposes) after being returned to full duty. The Marine will be given a minimum of 30 days and no more than 90 days to complete the PFT. This score will replace the PPFT taken during the waiver period. This is the only time a retest score will be used to replace a previous entry (partial) reported in MCTFS; this is not considered a remedial score.

3. Consecutive PPFTs. Marines who take PPFTs for two consecutive semi-annual periods, for the same injury or illness, will be referred to a MO and processed for a medical board. A MEB, PEB, or both will be conducted in order to determine the Marine's fitness for duty, per reference (i).

2203. ADMINISTRATIVE PROCEDURES

1. PFT/PPFT Recording

a. Unit Diary. Only when the command declares that the PFT scheduled is for the semi-annual requirement, will scores be reported by a Unit Diary entry and entered in the MCTFS, per reference (e). Only one PFT score will be reported on the Unit Diary per semi-annual period. No retest or remedial PFT scores will be recorded on the Unit Diary. PFTs taken for inventory purposes, such as a Marine reporting into a new unit or school, or taken for screening checklist purposes, will not be reported on the Unit Diary. In addition, PFT scores conducted for unit PT or for Remedial Physical Conditioning Program (RPCP) testing may not be used (or replace a score for) the semi-annual requirement.

b. FitRpt. All PFT score reporting will be done per reference (d), paragraph 4003.8b.

(1) Semi-annual PFT scores will be recorded in section A, "Administrative Information," under "Special Information" in block 8b.

(2) Mandatory comments will be made in section I, "Directed & Additional Comments," for the following circumstances:

(a) Superior Performance. Score of 285 and above.

(b) Not Medically Qualified (NMED). Limited duty status. Refer to paragraph 2300.3 for further details.

(c) PPFT. Sgts and above that have been granted a medical excuse by a MO, and who complete and pass a PPFT will have PART entered in block 8b, section A of their FitRpt. A brief explanation of the necessity for the PPFT, (e.g., TLD or PLD must be included in section I. A failed PPFT will be reported the same as for the PFT.

(d) PFT/PPFT Failure. A PFT/PPFT failure requires that Sgts and above receive an adverse FitRpt at the end of the next reporting period. In the case where a Marine does not have a reporting period until after the next semi-annual PFT, the previous PFT failure must be recorded on that FitRpt report, regardless of the performance on the more recent one.

(e) Required but Did Not Take (RDNT). A RDNT requires Sgts and above receive an adverse FitRpt. RDNT will be recorded in block 8b, section A and an explanation must be included in section I of the FitRpt.

(3) Height and weight measurements (and percent body fat, when necessary) will be transcribed from the PFT and Body Composition Worksheet for that reporting period.

c. Proficiency/Conduct (PRO/CON) Marks. Any Marine ranked Cpl or below, will receive commensurate PRO/CON marks for superior performance, PFT failure, or RDNT per reference (g), para 4005.

d. Page 11 SRB/OQR Entries. Page 11 entries are not required for Marines who fail a PFT/PPFT or for when a PFT/PPFT is RDNT.

2. PFT/PPFT Failures

a. Promotion/Reenlistment/Special Schools. Marines may be denied promotion, reenlistment, and/or special schools until they pass a PFT and fulfill all RPCP requirements. See paragraphs 1204 and 6009.2.e of reference (j).

b. Administrative Separation. Marines may be subject to administrative separation, per chapter 6 of reference (h), paragraphs 6105 and 6206, for the following circumstances:

(1) Unsatisfactory Performance. Repetitive PFT failures, due to lack of physical conditioning, displays poor self-discipline in a Marine's responsibility to maintain fitness standards.

(2) Medical Condition. Repetitive PPFT failures due to a medical condition.

3. Summary of Administrative Actions. Table 2-3 lists the administrative actions required for Marines who fail the PFT/PPFT.

ADMINISTRATIVE ACTION	NUMBER OF FAILURES	
	PFT 1st	PFT 2nd+
Mandatory RPCP	YES	YES
Page 11 SRB/OQR Mandatory Entry	NO	NO*
Adverse FitRpt or Substandard PRO/CON Marks	YES	YES
Eligible for Promotion	NO	NO
Eligible for PCS Transfer:	YES	YES
Process for Administrative Separation	NO	YES
May Be Denied Re-enlistment	YES	YES
May Be Denied Special Schools	YES	YES
May Be Denied Recommendation for Promotion	YES	YES

Table 2-3.--Administrative Actions.

*[Page 11 entries are required when Marines have failed consecutive PFTs or PPFTs, and when process for administrative separation is anticipated. See paragraph 6206, reference (h).]

CHAPTER 2

PHYSICAL FITNESS TEST (PFT)

SECTION 3: MEDICAL CONSIDERATIONS

2300. SCOPE. When a Marine is unable to take or pass the PFT due to a physical or medical condition, then a medical evaluation by a MO must be conducted, in order to excuse the Marine from participating in a particular event or the entire test. If the condition becomes extended or considered permanent in nature, then a determination by a formal MEB, a PEB, or both must be conducted to determine the Marine's medical qualification for continued active service. Marines evaluated for or assigned to a light duty or limited duty status are still required to perform those events of the PFT that are not affected by the condition, unless otherwise specified by the MO. A Marine may not be repeatedly excused from participation in the PFT/PPFT without a MEB action taken.

1. Light Duty. The status a MO assigns to a Marine when it is determined that a medical condition exists that temporarily interferes with the performance of their duty, but the Marine is expected to return to full duty (RTFD). A MO may recommend up to 30 days light duty. A Marine that is reevaluated and not returned to a full duty status in 30 days will be placed on medical hold, and a MEB must be initiated and completed within the following 30 days. Reference paragraph 2301 for further details on the MEB.

2. TLD. The status a MO assigns to a Marine when it is determined that a medical condition exists that requires an extended period of time before they can RTFD.

a. Marines, who are on TLD and are medically exempt from two consecutive PFTs or PPFTs for the same injury or illness, will be referred to a MO and processed for a MEB, a PEB, or both to determine fitness for duty.

b. A MEB must be conducted in order to assign a Marine to a TLD status. A Marine can be assigned to a TLD status for no more than 1-8 months by a medical treatment facility (MTF).

c. A PEB must be conducted in order to grant a Marine an extension of TLD, to be assigned a PLD status, or to be processed for separation.

3. Post-Light or Limited Duty PFT. Marines removed from light or TLD status by a MO, who did not take a PFT/PPFT for the semi-annual requirement, will be administered a PFT. COs should be attentive to the fact that a Marine returned to full duty status will need to train at a moderate pace to return to the level of fitness experienced prior to the injury/illness. As a result, a Marine will be given no less than 30 days and no more than 90 days preparation after being returned to a full duty status, to complete the PFT. If the grace period crosses over into the next semi-annual interval, the PFT score will satisfy the later requirement and PFT MED (medically excused) will be entered into MCTFS for the earlier testing period.

2301. MEB. A MEB is a group of MO's at a local MTF convened for the purpose of identifying the physical and mental qualifications of referred Marines to RTFD. The MEB is charged with determining whether a Marine's limitation(s) preclude their RTFD within a reasonable period of time or at all. A MEB will evaluate and report on the diagnosis, prognosis, possible RTFD, plan for further treatment, and make medical recommendations for disposition. Reference appendix K of this Manual for further details on when a MEB is required.

1. Case Referrals. A MEB case referral will be made on Marines not returned to a full duty status after the maximum 30 days of light duty, or medically waived from completing a PFT or PPFT for two consecutive semi-annual periods for the same medical condition. A MEB may recommend:

a. Returning the Marine to full duty.

b. A period of TLD if the initial light duty period of 30 days has not rectified the problem.

c. Placing the Marine on medical hold. Medical hold is a temporary status, not to exceed 30 days, which is used to allow a MEB to convene and further evaluate a Marine's case.

d. Forwarding the case to CMC (MMSR-4) for review.

(1) The MTF must notify the Marine's command and must notify and forward a copy of the abbreviated limited duty form to CMC (MMSR-4) via naval message.

(2) Upon review, TLD may be approved, or the MEB may forward the case to the PEB for determination of fitness for continued military service.

e. Assigning the Marine to TLD. TLD is an assignment of limited duty for a specified period. Assignment to TLD can only be made after the completion of a MEB, and only if the prognosis is that the Marine can be RTFD within the specified period. The cumulative period of TLD will not exceed 16 months. The MCTFS duty limitation code for this status is "Q, LD MED BOARD" per paragraph 1206 of reference (i). The Marine's command is responsible for entering the code into MCTFS when TLD is authorized.

(1) Enlisted Assignment to TLD. A MEB recommending an initial period of TLD up to 8 months for enlisted Marines may be approved at the local MTF without approval from CMC (MMSR-4).

(2) Officer Assignment to TLD. A MEB recommending any period of TLD for Marine Officers must be submitted to CMC (MMSR-4) for review and approval.

(3) Medical Reevaluation. A MEB reassigned TLD period. The MEB at the local MTF must inform the command and CMC (MMSR-4) of the Marine's updated medical status prior to the completion of the TLD period.

(4) Assignment Extensions. Subsequent enlisted and officer assignment that extend the TLD status beyond 8 months must be submitted to CMC (MMSR-4) for review and approval.

f. Forwarding the case to a PEB for determination of fitness to continue military service. Chapter 8 of reference (h) and paragraph 2302 below further detail actions to be taken.

2302. PEB

1. The PEB acts on behalf of the SecNav to make determinations of fitness for continued military service, entitlement to benefits, disability ratings and disposition of the medical case. A PEB is required to convene for cases that exceed the maximum 16 months of limited duty or when it is not anticipated that a Marine will RTFD consequent to the initial diagnosis and prognosis. Reference appendix K of this Manual for further details on when a PEB is required.

2. PEB Findings

a. "Fit" for Continued Naval Service. To be found "Fit" to continue Naval service by the PEB, evidence must exist to demonstrate that the Marine is reasonably able to perform (without medical impairment) the duties of their billet, grade, or rank remaining on their period of service obligation to include Reserve obligation. Marines found "Fit" by the PEB are eligible for appropriate assignment. A finding of "Fit" does not preclude subsequent determinations of unsuitability for deployment, PFT participation, disqualification for special duties, TLD, or administrative action (including separation) resulting from such determination.

b. "Unfit" for Full Duty. To be found "Unfit" by the PEB to continue Naval service, evidence must exist to demonstrate that the Marine is unable to reasonably perform the duties of their billet, grade, or rank remaining on their period of service obligation to include Reserve obligation. The PEB will direct CMC (MMSR-4) to separate or retire Marines found "Unfit" to continue Naval service.

c. PLD. PLD is a continuation on active duty in a PLD status, which is approved after a Marine has been found "Unfit" by a PEB as a result of physical disability. PLD is assigned when a Marine is not expected to RTFD earlier than 16 months or it is projected that the condition cannot be corrected. A Marine can remain on active duty in a PLD status, provided the criteria (e.g., $\geq$ 18 and $\leq$ 20 years active duty) is met per reference (d). CMC (MMSR-4) may screen, review, and grant requests for PLD not meeting the mentioned criteria on a case-by-case basis.

(1) SRB/OQR Entry Requirements. When a permanent medical condition has been identified, appropriate entries will be placed in the Marine's health record and SRB/OQR. This will eliminate the necessity to repeatedly issue medical waivers. The health record entry will include MEB/PEB findings, including the date of the medical determination, diagnosis, prognosis, and restrictions imposed. The NAVMED 6100/5, Abbreviated Limited Duty Medical Board Report, will be filed on the document side of the SRB/OQR after completing the MCTFS entry requirements.

(2) MCTFS Entry. PLD is requested through the PEB and approved by CMC (MMSR-4). PLD requests will be authorized by the PEB based on the best interest of the Marine Corps and the health of the Marine. The MCTFS duty limitation code for this status is "S, RETN LD STAT" per paragraph 1206 of reference (i). Only CMC (MMSR-4) is authorized to enter or remove this code from MCTFS.

(3) FitRpt Requirement. Marines on limited duty for an entire semi-annual period will have "not medically qualified (NMED)" recorded in block 8b of section A and an explanation included in section I of the FitRpt per reference (d).

(4) A Marine assigned to PLD for a period greater than 6 months, must be reevaluated and have a complete MEB sent to the PEB no later than 4 months prior to PLD expiration date. A copy will be forwarded to the CMC (MMSR-4). This permanent condition will be reevaluated and updated at each service-related required physical evaluation (e.g., annual, reenlistment, etc.).

2303. BODY COMPOSITION STANDARDS AND LIGHT OR TLD. Marines assigned a light or TLD status or placed on medical hold for MEB/PEB screening will maintain body composition standards. Being screened for or placed on light or TLD does not absolve the Marine's responsibility to comply with Marine Corps standards. A Marine that fails to maintain body composition standards may be placed on a BCP, assigned to a modified RPCP (reflecting physical restrictions as determined by a MO), and given dietary adjustments to follow. Light or TLD Marines assigned to a BCP, who fail to successfully return to Marine Corps standards upon completion of the Program, are subject to administrative action to include a Page 11 counseling entry for unsatisfactory performance per paragraph 6105, and processed for administrative separation per reference (h).

2304. REELISTMENT/PROMOTION

1. Enlisted

a. Reenlistment. Enlisted Marines shall not be granted reenlistment while assigned to a PLD status unless the CMC grants a specific waiver. Instead, the Marine should be placed on medical hold, Convenience of the Government Medical (COFGM), until found "Fit" or "Unfit" by a PEB per paragraph 8607 of reference (h).

b. Promotion (TLD). Enlisted Marines assigned a TLD status do not lose their eligibility to be considered for promotion, due to medical conditions diagnosed by a MO that limits full duty status. Refer to paragraph 8608.8 of reference (k).

c. Promotion (PLD). Enlisted Marines who have been previously selected and are pending a fitness for duty determination by the Disability Evaluation System (DES) will be presented their certificates of appointment, and their promotions will be affected, if otherwise qualified. However, enlisted Marines assigned to a PLD status on the convening date of the appropriate selection board are not eligible for consideration for promotion.

2. Officers. If an officer is in a PLD status, delays of promotion can be initiated at the time an officer is selected by a selection board or prior to the time an officer's promotion is to be effected. Delay of an officer's promotion is warranted if the SecNav, CMC, or CO believes the officer is mentally, physically, morally, or professionally not qualified for promotion to the next higher grade. Refer to paragraph 8002.9 of reference (h) and paragraph 6003 of reference (m).

2305. PREGNANCY/POSTPARTUM. After medical determination of pregnancy, Marines will be waived from the PFT. Pregnant Marines will participate in a medically approved exercise program throughout the pregnancy and postpartum period, except when not recommended by a MO.

1. Resuming Physical Conditioning and Testing. Postpartum Marines will be required to take and pass a PFT no later than 6 months from the date that a MO has authorized her RTFD. This full duty authorization will normally coincide with the end of the 6-week (postpartum) convalescent leave period, per reference (l). Unless medically prohibited, physical conditioning should resume at this time. COs should be attentive to the fact that a Marine RTFD status will need to train at a moderate pace to return to the level of fitness she experienced prior to her pregnancy.

2. Early Termination/Miscarriage. A Marine whose pregnancy results in a miscarriage or has been terminated will be required to conform to body composition standards and pass a semi-annual PFT, in a time period determined by a MO. Just as with each pregnancy, recovery periods will vary depending on the individual's physical status at the time of termination/miscarriage. Time limits will not exceed those that are prescribed for a full term pregnancy.

3. Difficult Pregnancy or Birth/Cesarean Section/Still Birth. Although rare, these types of instances may result in a more difficult recovery. Therefore, in pregnancies where complications develop, MOs will determine on a case-by-case basis the time needed to return a Marine to full duty. Upon RTFD, the Marine will be required to conform to the body composition standards and pass the semi-annual PFT in the time period determined by a MO.

4. Lines of Communication. It is important that an open dialogue and coordination between the Marine, the MO, and the CO be maintained in order to determine any physical limitations and dietary concerns, the healing time required to return the Marine to full duty, and the Marine's overall progress.

5. Failure to RTFD. Marines not RTFD at the termination of the 6-week (postpartum) convalescent leave period, may be given an additional period of up to 30 days light duty, to allow additional recovery time. If upon termination of the 30 days light duty, the Marine is still not RTFD, then the Marine will be placed on medical hold and referred to a MEB for determination of fitness for continued military service. A Marine not RTFD status and as a result referred to a MEB will not be assigned to the BCP for failure to meet weight and body fat standards. Chapter 2, paragraph 2301 further details the MEB process and appendix L describes the Pregnancy and (inactive) BCP Assignment Process.

6. Promotion and Reenlistment

a. Pregnant Marines who did meet the reenlistment or promotion criteria prior to medical determination of pregnancy are eligible to reenlist and be promoted at any time during their pregnancy or (postpartum) convalescent period.

b. Pregnant Marines who did not meet reenlistment or promotion criteria prior to medical determination of pregnancy will not be eligible to reenlist or be promoted during the pregnancy period. Reenlistment and promotion eligibility will occur for these Marines only after they have been RTFD and have met all criteria.

2306. MEDICAL WAIVERS/EXEMPTIONS

1. Medical Waivers. A medical waiver (light or TLD) that excuses a Marine from participating in a PFT will not be considered a PFT failure. The MO will document on the medical evaluation the medical condition, the expiration date, physical exercise (based on limitations), and the PPFT events that the Marine can complete, if any, during the waiver period. Marines who are medically excused from performing a particular event of the PFT, as certified by a MO, are required to perform a PPFT. Reference paragraph 2202 of this Manual for PPFT requirements.

2. Injury During PFT. Marines injured during the execution of a semi-annual PFT (e.g., pulled muscle during the run) may be given credit for a PPFT score if two of the three events were completed prior to the injury. Or, after clearance from a MO, the Marine will be given an opportunity to run the PFT again. Only the CO and the Command PT Representative, after reviewing the facts surrounding the incident, can authorize that credit be given for a PPFT or if a Marine is given an opportunity to run the PFT at a later time. Marines completing only one event

prior to the injury will receive neither a PFT score nor PPFT credit. If evaluated and assigned a light duty status as a result of the injury, then the Marine will be required to retake the PFT 30-90 days after being RTFD status. Marines who are evaluated and not assigned a light duty status, will retake the PFT at the CO's discretion.

3. Illness. Marines encountering a rapid onset illness prior to the semi-annual PFT will obtain a light duty chit. Light duty chits must be submitted to the command prior to the scheduled PFT. The CO will decide when the Marine will be required to take the missed PFT.

2307. AGE 46 AND OVER RISK FACTOR SCREENING

>CH 1 1. Marines age 46 years and over will self-validate a "Risk Factor Screening" as a precautionary measure, 30-60 days prior to each semi-annual PFT. However, during IGMC/CGI team inspections, the inspectors may direct personnel age 46 and over to complete the Risk Factor Screening checklist if one is not on file. Marines will complete sections (A) and (B) of the Risk Factor Screening contained in appendix E, and submit the completed form to their CO for review. A medical screening by a MO, prior to PFT/PPFT participation, will only be required if the Marine:

a. Answers "yes" to any question in section (A) or (B).

b. Exceeds the body composition standards.

2. Marines requiring a follow-on medical screening must be medically cleared by a MO prior to participating in the PFT. Supplemental SF600 forms will be attached to the Screening if additional MO remarks are required. Independent Duty Corpsmen are not authorized to issue medical clearance to Marines who answered "yes" to any risk factor screening question. Marines who are not cleared by a MO to participate in the semi-annual PFT are restricted from participation during the current semi-annual period, either voluntarily or involuntarily. Such restriction only applies to the semi-annual period during which the MO determination was made.

3. Marines will participate in the PFT only if cleared by a MO after the referral, and no further medical evaluation is required. The signed and completed Risk Factor Screening (original) and SF600 documents will be filed in their health records, and a copy of the screening will be given to the CO. The Command PT Representative will annotate the Marine has completed the screening (been cleared for participation), on the PFT and Body Composition Worksheet.

4. Marines having a current physical examination or Navy Annual Health Evaluation (BUMEDINST 6120.26) will not be required to complete the semi-annual Risk Factor Screening or receive additional medical clearance to participate in the PFT if:

a. Examination or evaluation was conducted within the last 6 months from the date of the scheduled PFT.

b. There were no physical restrictions or limitations imposed by the MO.

c. Their state of health has not since changed.

CHAPTER 2

PHYSICAL FITNESS TEST (PFT)

SECTION 4: REMEDIAL PROCEDURES

2400. RPCP

1. The command directed RPCP is designed to improve a Marine's physical fitness and body composition, through command-structured physical conditioning, guidance, and supervision as contained in appendix C.

2. Assignment. Marines will be placed on a RPCP for any of the following:

a. Failing a semi-annual PFT.

b. Marines assigned to the BCP.

c. Marines required to take a semi-annual PFT but fail to do so ("required did not take, RDNT").

d. Marines who consistently demonstrate poor physical performance in MOS related tasks.

e. Female Marines assigned to the BCP prior to pregnancy and RTFD status will resume active participation in the RPCP. The level of program participation may require MO input, if physical limitations are identified.

3. Program Requirements. RPCP requirements:

a. Minimum of 30 days in the Program, regardless of change in ability.

b. Take and pass a PFT at completion of the initial 30-day period.

c. Marines assigned to the BCP will remain in the RPCP for the duration of the BCP assignment.

d. No formal administrative action is required for assignment to RPCP (e.g., no requirement for a SRB/OQR, Unit Diary, or MCTFS entry).

CHAPTER 3

BODY COMPOSITION PROGRAM (BCP)

CHAPTER 3

BODY COMPOSITION PROGRAM (BCP)

3000. SCOPE. It is essential that every Marine maintain the established standards of health and physical fitness. With age, it is not uncommon for the average American to experience a decrease in physical activity, a change in metabolic rate, with unchanged caloric intake. As a result, body fat and weight may increase, having a negative impact on one's health and level of fitness. Medical authorities state that excess body fat can be associated with high blood pressure, high blood cholesterol, diabetes, cancer, cardiovascular disease, and similar health risks. In extreme climates and physically demanding environments, excess weight and body fat can also impede physical performance and stamina, which is of particular importance to the Marine and the mission.

3001. OBJECTIVES. The objectives of the Marine Corps' BCP are:

1. To establish healthy weight and body fat standards that ensure all Marines are physically capable to perform their duties.

2. To contribute to the health and well being of every Marine by continuously monitoring weight, body fat, diet/nutrition, and physical fitness conditioning.

3. To motivate all Marines to set the example by maintaining the established body composition standards.

4. To ensure those Marines who do not meet established standards are counseled accordingly, and given the opportunity and proper guidance to achieve the standards through the BCP and Remedial Physical Conditioning Program (RPCP).

3002. RESPONSIBILITIES

1. Individual Marine. It is every Marine's responsibility to maintain the Marine Corps' body composition standards, or if not in adherence with standards, to take appropriate action to return to them in a timely manner. Marines on light or limited duty, whose medical condition precludes them from participating in specific activities, will be expected to participate in conditioning alternatives and dietary adjustments, in order to maintain these standards.

2. CO. It is the CO's responsibility to ensure each Marine receives a semi-annual Body Composition Evaluation, and to closely monitor Marines assigned to the command's BCP and RPCP. COs will:

a. Continually monitor all Marines within the command to ensure adherence to established body composition standards.

b. Identify Marines who exceed the weight and body fat standards by conducting periodic Body Composition Evaluations.

c. Offer educational programs, events, and other motivational means to encourage Marines to achieve and maintain appropriate body composition standards. Such programs could include, but are not limited to: body composition seminars, health and fitness fairs, and individual consultations conducted by qualified dieticians, medical authorities, and Semper Fit personnel. Spouses should be encouraged to participate in these programs with the assigned Marine. Reference the Semper Fit website (www.usmc-mccs.org) for additional information and services.

d. Ensure the leadership understands that Marines may not have the education and experience needed to attain and maintain proper health and fitness, and that lifestyle changes may take time. However, if given the proper tools (e.g., education, supervision, and positive reinforcement), leaders can assist Marines in achieving and maintaining a healthy and fit lifestyle more quickly.

3. Medical Officer (MO). The MO's responsibility is to evaluate the Marine's body composition, overall health, diet, and physical ability to participate in the BCP and RPCP. The CO should receive the following from the MO:

a. Determination of whether the weight or body fat increase is due to an underlying cause or associated disease.

b. The Marine's current duty status (e.g., full, light, or limited duty).

c. The recommended dietary plan and weight and body fat reduction goals.

d. The physical conditioning restrictions or limitations imposed.

e. The date the Marine is expected to return to full duty (RTFD) status.

f. Any other pertinent information that pertains to the Marine's individual case.

CHAPTER 3

BODY COMPOSITION PROGRAM (BCP)

SECTION 1: STANDARDS

3100. BODY COMPOSITION STANDARDS. The Marine Corps' weight and body fat standards are health and performance based, and not based on appearance. Marines are considered not within these standards when their body weight and body fat exceed the maximum limits as contained in appendix I. If tested and the Marine's percent body fat exceeds the maximum limit (18% for males, 26% for females), then the CO will evaluate the Marine's physical performance (e.g., current semi-annual Physical Fitness Test (PFT)) using the Physical Performance Evaluation in paragraph 3102. If the Marine does not meet the Physical Performance Evaluation criteria, then the CO will have the Marine evaluated by a MO and assigned to a BCP - formerly known as the "Weight Control Program." Marines assigned to the BCP will receive assistance in reducing body weight and in particular body fat, in order to attain and maintain a more healthy physical fitness state. The BCP assignment process and endorsements are contained in paragraph 3201 and appendix D of this Manual.

3101. BODY COMPOSITION EVALUATION PROCEDURES

>CH 1 1. Body Composition Evaluation (BCE). The BCE is a semi-annual (January-June and July-December) requirement (annual for Reserves) designed to ensure Marines are within HT/WT or BF standards. However, commanders may direct BCE's for a unit or individuals as often as deemed necessary. BCE's may be conducted on the same day as the PFT. IGMC/CGI teams may also conduct unannounced BCE's in support of inspections.

a. Weigh-Ins. Regardless of duty status (full/light/limited duty), all Marines are required (at a minimum) to complete a semi-annual (annual for Reserves) Body Composition Evaluation as prescribed by the unit training schedule. The Command Physical Training (PT) Representative and his/her designate are the **only** unit personnel authorized to conduct Body Composition Evaluations, and will record the results on the PFT and Body Composition Worksheet in appendix F.

(1) Uniform. Uniform for the Evaluation is the USMC established green-on-green T-shirt and shorts, and socks. No other uniform or PT attire is permitted. Shoes will not be worn.

(2) Height Measurement. Height will be measured using a standard, non-stretching (cloth or fiberglass) tape measure that is secured to a wall and flush with the deck; weight scales that have a height measuring device will not be used. The Marine will stand with their back against the wall, heels flat on the deck, shoulders back, with arms to the side in a relaxed manner, and head straight forward. A ruler will be used to mark the Marine's exact height.

(a) Measurement will be recorded to the nearest inch.

(b) If the height fraction is less than 1/2 inch, round down to the nearest inch. If height fraction is 1/2 inch or more, round up.

(3) Weight Measurement. Weight will be measured on a calibrated balance beam scale in the required PT uniform, without running shoes. A one-pound reduction will be granted for the PT uniform; however, no other weight reductions are authorized.

(a) Measurement will be recorded to the nearest pound.

(b) If the weight fraction is 1/2 pound or less, round down to the nearest pound. If more than 1/2 pound, round up.

(4) Refer to section I of appendix I to review the height and weight chart. If a Marine is not within height and weight standards, then the Marine will be measured for body fat immediately (same day). Body fat estimation is only authorized for Marines who exceed their maximum weight standard. The height and weight measurement conducted during this portion of the Body Composition Evaluation is what will be reported on the FitRpt.

b. Body Fat Estimation. Body fat will be estimated using the circumference-based method with one set of measurements for each gender (Males: height, neck circumference, and abdominal circumference at the navel. Females: height, neck circumference, waist circumference at the thinnest portion of the abdomen, and hips). No substitute methods of assessment are permitted. This DoD directed method of body fat estimation has been carefully evaluated for applicability to service members and represents the best approach that can be applied with minimal error (+/- 3 to 4%). This method is also valid because of the emphasis on abdominal circumference, the site of human body fat deposition which is most strongly associated with health risks, and which corresponds to other military goals including appropriate appearance and healthy exercise habits. Body fat measurements will be taken in the following manner:

(1) Marines will report for body fat estimation in the required PT uniform. The Command PT Representative and his/her designate will conduct the measurements.

(2) The tape measure used should be made of non-stretchable material, preferably fiberglass; cloth or steel tapes are unacceptable. The tape measure width should be 1/4 - 3/8 inch. The tape measure should be calibrated (i.e., compared with a yardstick or a metal ruler to ensure validity). This is done by aligning the fiberglass tape measure with the quarter inch markings on the ruler. The markings should match those on the ruler; if not, do not use that tape measure.

(3) All measurements will be taken on bare skin with the only exception being the female hip measurement.

(4) For use in the body fat estimation formula, Marines will have their height re-measured and recorded to the nearest 1/2-inch.

(a) If the measured height fraction is greater than an inch, round up to the nearest 1/2-inch.

(b) If the measured height fraction is greater than 1/2-inch, round up to the next whole inch.

(c) If the measured height fraction is at 1/2 inch, use that measurement in the body fat estimation equation.

(d) The height measured at the initial semi-annual (annual for Reserves) height/weight screen is what will be reported on the FitRpt, not the height used for the body fat estimation formula.

(5) Without causing indentation, the tape will be applied to the skin with sufficient tension to hold it in place as the measurement is being read.

(6) Circumference measurements will be taken three times to ensure accuracy. Each set of measurements will be completed sequentially to discourage assumption of repeated measurement readings. For males, complete one set of abdomen and neck measurements, **not**, for example, three abdominal circumferences followed by three neck circumferences. Continue the process by measuring the abdomen and neck in series until three sets of measurements are completed. For females, complete one set of waist, hip, and neck, measurements, **not**, for example, three waist measurements followed by three hip measurements. Continue the process by measuring the waist, hip, neck series three sets of measurements are completed. The lower of the three body fat measurements will be recorded on the PFT/Body Composition Worksheet in appendix F.

(7) Procedures for Body Fat Estimation for Males:

(a) Measure the neck circumference by placing the edge of the tape measure flush with the bottom of the larynx and perpendicular to the long axis of the neck. The Marine should look straight ahead during the measurement, with shoulders down (not hunched). For neck measurements in excess of the whole inch, round the neck measurement up to the nearest 1/2-inch and record (e.g., round up 16 1/4 inches to 16 1/2 inches).

(b) Measure abdominal circumference against the skin at the navel, level and parallel to the deck. Arms are at the sides. Take measurement at the end of the Marine's normal, relaxed exhalation. Round the abdominal measurement down to the nearest 1/2-inch and record (e.g., round down 34 3/4 to 34 1/2 inches).

(c) Determine percent body fat by subtracting the neck from the abdominal measurement and comparing this value against the height measurement, (e.g., abdominal - neck = circumference value). Refer to the chart "Percent Body Fat Estimation for Males" in section II of appendix I.

(8) Procedures for Body Fat Estimation for Females:

(a) Measure the neck circumference by placing the edge of the tape measure flush with the bottom of the larynx and perpendicular to the long axis of the neck. The Marine should look straight ahead during the measurement, with shoulders down (not hunched). To establish neck measurements in excess of the whole inch, round the neck measurement up to the nearest 1/2-inch and record (e.g., round up 13 3/8 inches to 13 1/2 inches).

(b) Measure the natural waist circumference against the skin at the point of minimal abdominal circumference, usually located about halfway between the navel and the lower end of the sternum (breastbone). When this site is not easily observed, take several measurements at probable sites and use the smallest value. Ensure the tape is level and parallel to the deck. Arms are at the sides. Take measurements at the end of a normal, relaxed exhalation. Round natural waist measurement down to the nearest 1/2-inch and record (e.g., round down 28 5/8 inches to 28 1/2 inches).

(c) Measure the hip circumference while facing the Marine's right side by placing the tape around the hips so that it passes over the greatest protrusion of the buttocks as viewed from the side. Ensure the tape is level and parallel to the deck. Apply sufficient tension on the tape to minimize the effect of clothing. Round the hip measurement down to the nearest 1/2-inch and record (e.g., round down 38 3/8 inches to 38 inches).

(d) Determine percent body fat by adding the waist and the hip measurements, subtracting the neck measurement, and comparing values against the Marine's height measurement (e.g., waist + hip - neck = circumference value). Refer to the chart "Percent Body Fat Estimation for Females" in section II of appendix I.

(e) Marines who meet the body fat standards, 18% for males and 26% for females, will be considered within the Marine Corps' body composition standards. Marines who exceed the body fat standards will be further evaluated on criteria set forth in paragraph 3102 below.

3102. PHYSICAL PERFORMANCE EVALUATION. There is an integral relationship between body composition and physical performance that warrants consideration when evaluating the "total" Marine. For instance, there are Marines who exceed weight and body fat standards, yet perform exceptionally well on the semi-annual PFT and combat fitness tests. These Marines demonstrate both exceptional cardiovascular endurance and upper body strength, with a body composition that may be identified as slightly exceeding the standards, but overall could be viewed as maintaining a reasonable body weight. In such cases, the Marine Corps' view on physical fitness as "essential to the day-to-day effectiveness and combat readiness" should be the overriding factor in these circumstances. The Physical Performance Evaluation not only takes body composition and physical fitness performance into consideration, it also takes into account the acknowledged variance in the body fat estimation technique ("taping" method). The Physical Performance Evaluation accommodates for the acknowledged margin of error in body fat estimation and testing, normally a variance of 3 to 4%. Reference (f) establishes a requirement to make such an accommodation in view of this variance. In essence, the Physical Performance Evaluation assesses the "total" Marine (e.g., body composition and recent semi-annual PFT performance as well as acknowledges the margin of error in body fat testing).

1. The Physical Performance Evaluation requires a Marine to meet the following criteria:

>CH 1 a. A 1st Class PFT taken a maximum of 90 days before or 30 days after the body composition evaluation.

b. Body fat estimation does not exceed the standard by more than 4%, ($\leq$22% for males, $\leq$30% for females).

2. Marines exceeding 22 and 30% respectively or score a 2nd Class PFT or lower will not be eligible for the Physical Performance Evaluation under any circumstance. These Marines will automatically be processed for a BCP and RPCP assignment.

3. Partial Physical Fitness Tests (PPFT) will not be considered. Marines must be fit to take a PFT and score a 1st Class to be considered for the Physical Performance Evaluation.

4. No adverse administrative action will be taken for Marines meeting the Physical Performance Evaluation criteria. The Body Fat estimation percentage will still be reported on the FitRpt. The following is a directed comment in section I on the FitRpt meeting the Physical Performance Evaluation criteria:

> "MRO meets the Physical Performance Evaluation criteria in MCO P6100.12 and is within standards."

5. Marines meeting the Physical Performance Evaluation criteria will only be exempt from administrative action for a particular 6-month (semi-annual) period (or annual for Reserves). Marines will be reevaluated at the next semi-annual PFT (annual for Reserves) and Body Composition Evaluation per paragraph 2100.

6. Physical Performance Evaluation examples:

a. Example 1: A male Marine, who exceeds his maximum allowable weight, is measured at 21% body fat, and scores a current 1st Class PFT has met the Physical Performance Evaluation criteria and would not be assigned to the BCP.

b. Example 2: A male Marine, who exceeds his maximum allowable weight, scores a current 1st Class PFT, but is measured at 24% body fat, does not meet the Physical Performance Evaluation body fat criteria, and therefore would be assigned to the BCP.

c. Example 3: A female Marine who exceeds her maximum allowable weight, is measured at 32% body fat, and scores a current 2nd Class PFT, does not meet the Physical Performance Evaluation body fat and PFT criteria, and therefore would be assigned to the BCP.

3103. PREGNANCY/POSTPARTUM BODY COMPOSITION REQUIREMENTS. Marines will be exempt from weight and body fat standards during pregnancy, but will maintain active participation in the Physical Conditioning Program (PCP) unless otherwise prescribed by her MO. Marines in a postpartum status will be required to meet weight and body fat standards no later than 6 months following a MO's authorization to RTFD. This authorization normally coincides with the completion of the 6-week (postpartum) convalescence leave period, per reference (l).

1. Miscarriage/Stillbirth. A Marine whose pregnancy terminates prematurely or that results in a stillbirth will meet weight and body fat standards within a time period determined by a MO. Due to the varying disposition of this medical situation, it is important that an open dialogue exists between the Marine, the MO and her CO to determine the time needed to recover and conform to body composition standards. The recovery period in this case, however, will not exceed the 6 months authorized for full-term pregnancies.

2. BCP Assignment. Marines who become pregnant while assigned to the BCP will remain on the Program in an "inactive" status.

a. The only Unit Diary entry required when a Marine becomes pregnant while assigned to a BCP is to report MCTFS Duty Limitation Code of "N, PREGNANCY" per paragraph 1206 of reference (i). An advisory message will be generated on the unit's Diary Feedback Report (DFR) as follows: "MARINE PLACED IN INACTIVE WT CNTL STATUS." All further advisories concerning the Marine's status in the BCP will be suppressed for the duration of the pregnancy and the medical related periods of limited duty immediately following the pregnancy.

b. Following the MO's medical evaluation, the Marine will be returned to either a full duty status, or reported to a non-medical related limited duty status. The Marine will then resume active participation in the BCP. All Marines will be assigned for a 6-month period, regardless of time spent on the Program prior to pregnancy. The unit will receive the following advisory message on their DFR: "MARINE RETURNED TO ACTIVE WT CNTL STATUS FOR SIX MONTHS."

3. Promotion Eligibility. Marines will not be eligible for promotion if they are in an "inactive" status for a BCP assignment per paragraphs 1204.3n and 6009.2e of reference (k). Only when the Marine returns to the weight and body fat standards and meets all other promotion requirements will she be promoted. Reference paragraph 2305 of this Manual for further details.

CHAPTER 3

BODY COMPOSITION PROGRAM (BCP)

SECTION 2: BCP PROCEDURES

3200. BODY COMPOSITION MEDICAL EVALUATION

1. Marines not within the body composition standards and do not meet the criteria for Physical Performance Evaluation will be referred to a MO to be evaluated. A medical evaluation and clearance, and all the Program endorsements contained in appendix D must be completed before a Marine can be assigned to a BCP. During the MO's evaluation, a determination must be made on whether or not a Marine's increase in weight and body fat is due to an underlying cause or associated disease, as follows:

>CH 1 a. Underlying Cause Defined. The following conditions may be considered as an underlying cause or associated disease which exempts a Marine (temporary basis) from meeting height/weight (HT/WT) or bodyfat (BF) standards and not being assigned to a BCP:

(1) Metabolic conditions, such as hypothyroidism, diagnosed by a medical officer (MO). The temporary exclusion will expire within 6 months of the metabolic condition being normalized through treatment or a limited duty/permanent disability board will be initiated.

(2) Medical conditions which require administration of medications that directly result in temporary weight gain.

>CH 1 b. Neuromusculoskeletal (NMS) Conditions. NMS conditions (e.g., sprained ankle, low back strain) that cause Marines to be placed in a not fit for full duty status may not be considered as an underlying cause or associated disease which exempts a Marine (temporary basis) from meeting HT/WT or BF standards. A NMS condition/not fit for full duty status determination also does not preclude assignment to a BCP. Assignment to a BCP for Marines in a not fit for full duty status secondary to a NMS condition will include referral from a MO or appropriately credentialed health care provider (ACHCP) to an appropriate agency for nutritional guidance and commencement of an alternative conditioning program.

>CH 1 c. Out of Standards Without Underlying Cause. When receiving the initial evaluation by the MO it is determined that no underlying cause or associated disease is the reason for the Marine's weight and body fat gain, then the Marine will continue forward with the BCP assignment process. The following medical recommendations will be made per an individual Marine's case:

(1) Recommend a realistic weight loss (e.g., pounds per week) and percent body fat reduction time line.

(2) Recommend caloric intake reduction and nutritional guidance (dietary measures).

(3) Recommend alternative physical conditioning exercises based on physical limitations (if necessary) for Marines on light or limited duty.

>CH 1 d. Out of Standards With an Underlying Cause. When receiving their initial evaluation by the MO, it is determined that an underlying or associated disease is the reason for the Marine's weight and body fat gain, then it must be thoroughly

annotated in the Marine's Health Record and on the Program's First Endorsement. The following courses of action may be taken per individual Marine's case:

(1) Treatment to alleviate the medical condition and return of the Marine to the command.

(2) Hospitalization for obesity treatment at an Alcohol Rehabilitation Center (ARC). Criteria for ARC Residential Obesity Program (ROP) are contained in paragraph 3202 in this Manual.

(3) Diagnosed as untreatable or permanent in nature, case forwarded to a PEB for determination of fitness to continue military service.

(4) RTFD, assignment to BCP and RPCP.

3201. BCP ASSIGNMENT. The following actions will be taken by the command when a Marine is being screened for a BCP assignment. Marines assigned to the BCP will be required to participate in the Program for a minimum of 6 months.

1. Preliminary Medical Evaluation. The CO will complete the BCP Evaluation Letter (appendix D, enclosure 1) requesting a preliminary medical evaluation. This Letter and the First Endorsement (appendix D, enclosure 2) will be forwarded to the MO at the local Medical Treatment Facility (MTF). The Marine will report directly to the MTF to be medically evaluated by the MO.

2. First Endorsement. The MO will complete the First Endorsement (appendix D, enclosure 2) and return the Letter to the CO.

a. The MO will attach the following information to the Endorsement:

(1) Determination of whether the weight and body fat increase is due to an underlying cause or associated disease;

(2) Recommended weight and body fat reduction goals;

(3) Recommended dietary measures;

(4) Physical limitations or restrictions imposed; and

(5) Medical clearance to participate in the Command's BCP and RPCP.

b. If the Marine is diagnosed with an underlying cause or associated disease, then the MO may recommend that the Marine not be assigned to a BCP until the Marine's condition is treated, or may recommend that the Marine be processed for medical separation for untreatable conditions. This diagnosis must be thoroughly annotated in the Marine's Health Record and on the Program's First Endorsement. Reference paragraph 3200 in this Manual for further details regarding "underlying cause and/or associated disease."

c. The signature block on the Endorsement will clearly state the name, grade, Branch of Service, medical title, and organization (as appropriate) of the MO who diagnosed the cause of the Marine's deficiency.

3. Second Endorsement. Upon reviewing the MO's diagnosis and recommendations, the CO will complete the Second Endorsement (appendix D, enclosure 3). The CO will counsel the Marine on the deficiencies, corrective actions required, and information regarding the BCP and RPCP, and will have a 6105 counseling entry recorded on the Marine's SRB/OQR's Page 11 per reference (h).

4. Third Endorsement. Upon completion of the counseling, the Marine will sign and date the Third Endorsement (appendix D, enclosure 4) acknowledging the deficiencies, corrective actions required, and requirements for the 6-month BCP and RPCP assignments.

5. Admin Section Responsibilities. The CO will direct the Admin Section to:

a. On receipt of the Third Endorsement (appendix D, enclosure 4):

(1) Enter the BCP assignment on the Unit Diary and record the Unit Diary number and date on the bottom of the Third Endorsement. Forward this Endorsement to the Command PT Representative to be maintained in the Marine's BCP file.

NOTE: A BCP assignment will not be run on the Unit Diary until the Marine has been evaluated and cleared for participation by a MO, the appropriate Program endorsements have been completed, and the Marine has been properly counseled by the CO.

(2) Use the following code/statement to report assignment to BCP in MCTFS: TTC 337 000 TO WT CNTL.

b. When directed, record 6105 counseling entry on SRB/OQR Page 11.

c. When directed, submit a request for Marine Corps Institute (MCI) Course 3316, "Basic Nutrition."

d. Notify the Command PT Representative when the Admin Section has received a Diary Feedback Report (DFR) that identifies a Marine's BCP assignment has expired.

e. When directed, make an entry on the Unit Diary extending the Marine's initial (first) BCP assignment and enter the following code/statement per reference (e) into MCTFS: TTC 337 001 EXTEND WT CNTL. Record the Unit Diary number and date on the Sixth Endorsement (appendix D, enclosure 7) and forward it to the Command PT Representative to be maintained in the Marine's BCP file.

f. When directed, remove the Marine from the BCP, making an entry on the Unit Diary and the following code/statement per enclosure (e) into MCTFS: TTC 339 000 FR WT CNTL. Record the Unit Diary number and date on the Sixth Endorsement (appendix D, enclosure 7) and return it to the Command PT Representative to be maintained in the Marine's BCP file.

6. Command PT Representative Responsibilities. The CO will direct the Command PT Representative to:

a. On receipt of the Third Endorsement (appendix D, enclosure 4) from the Admin Section, create a BCP file for the Marine. The BCP file will contain all Marine's Program Endorsements, medical documentation, counseling, body composition testing results, training log, etc.

b. Immediately initiate the 6-month RPCP for a Marine medically qualified. A Marine in a light or limited duty status will be required to participate in an alternative RPCP, based on the physical restrictions and limitations imposed by the MO.

c. Conduct diet and nutrition classes for a Marine or set up appointments with the fitness and medical professionals at the local Semper Fit facilities or MTFs.

d. Order (via Admin Section) MCI Course 3316, "Basic Nutrition," for the Marine (whether Officer or Enlisted), and direct the Marine to complete the Course prior to BCP assignment expiration.

e. Take weight and body fat measurements once every 2 weeks (once a month SMCR), and document results in the Marine's BCP file.

f. Document in the Marine's BCP file the counseling conducted for those Marines who have displayed neither weight loss nor body fat reduction, and if it is a result of the Marine not adhering to the regimen prescribed by the MO and the command.

(1) A Marine not making satisfactory progress during the BCP assignment will receive a 6105 counseling entry on the SRB/OQR's Page 11 for failure to make satisfactory progress. This Page 11 entry will be given for unsatisfactory performance per paragraph 6206 in MCO P1900.16F, and is required for administrative separation.

(2) A Marine who makes progress, but fails to meet the weight and body fat standards at completion of a BCP assignment will receive a 6105 counseling entry on the SRB/OQR's page 11. This Page 11 entry will be given for failing to meet body composition standards per paragraph 6215 in MCO P1900.16F, and is required for administrative separation.

g. Maintain and update accordingly, the BCP file, (e.g., counseling, body composition test results, and RPCP training log entries).

>CH 1 h. Sgts and above who are assigned to the BCP will be administered an adverse FitRpt per reference (d). Cpls and below will receive commensurate PRO/CON marks per reference (g).

7. Fourth Endorsement. After a Marine has participated in the BCP for 4 months, the CO will complete the Fourth Endorsement (appendix D, enclosure 5) and will forward this endorsement and the Fifth Endorsement (appendix D, enclosure 6) to the MO at the local MTF. The Marine will report to the MTF to be reevaluated by the MO. (If at any time the Marine is failing to lose weight or body fat, the Marine should be evaluated earlier, in order to be reexamined for a possible underlying or associated disease.)

8. Fifth Endorsement. The MO will reevaluate the Marine, complete the Fifth Endorsement (appendix D, enclosure 6), and forward the endorsement to the CO with one of the following diagnoses/recommendations:

a. Underlying Cause or Associated Disease. The Marine has been diagnosed as having an underlying or associated disease, and the Marine will:

(1) Be placed in an "inactive status" (duty limit "N") for the BCP assignment until the Marine's condition is treated; or

(2) Be recommended for (medical) separation for untreatable conditions.

b. Reference paragraph 3200 for further guidance on cases involving an underlying cause or associated disease.

c. No Underlying Cause or Associated Disease and Making Progress. The Marine has been diagnosed as not having an underlying cause or associated disease and is making satisfactory progress towards weight and body fat reduction goals. Recommend that participation in the BCP (and RPCP) continue until assignment expires.

d. No Underlying Cause or Associated Disease - Not Making Progress. The Marine has been diagnosed as not having an underlying cause or associated disease and is not making satisfactory progress towards weight and body fat reduction goals. Recommend that participation in the BCP (and RPCP) continue until assignment expires.

(1) A Marine may not receive an extension if lack of satisfactory progress is a result of not adhering to the regimen prescribed by the MO and the command.

(2) A 6105 counseling entry on the Marine's SRB/OQR's Page 11 will be administered at this time (fourth month) for failure to make satisfactory progress. See paragraph 6206 in MCO P1900.16F.

e. Returned to Weight or Body Fat Standards. The Marine has met weight and/or body fat reductions goals and is within body composition standards. Recommended that participation in the BCP (and RPCP) continue until assignment expires.

9. Expiration of Program Assignment. The Admin Section will receive a DFR when the BCP assignment has expired. One of the following actions will be taken:

a. A Marine has met and maintained the weight and body fat standards and will be officially removed from the BCP (and RPCP); see paragraph 13 in this section for Program removal instructions.

b. A Marine has failed to return to the weight and body fat standards, but has made satisfactory progress. Refer to paragraph 11 below for further guidance on Program (first assignment) extensions.

c. A Marine has failed to return to the weight and body fat standards and has not made satisfactory progress. CO will administer a 6105 counseling entry on the Marine's SRB/OQR's Page 11, per paragraph 6206 or 6215 in MCO P1900.16F respectively, and will process the Marine for administrative separation.

10. Extensions to First Assignment. A Marine making satisfactory progress and reasonably close to meeting their weight and body fat standards at the end of the 6-month BCP assignment may be granted a Program extension.

a. CO can grant a one-time extension of up to 6 months to first assignments only. An extension to a first assignment does not constitute a second assignment.

b. A Marine granted an extension will continue participation in the BCP and RPCP until the extension period expires. A Marine granted an extension will be counseled pursuant to paragraph 6105 of reference (h) with standard language in subparagraph (1) below and have the following code statement entered into MCTFS: TTC 337 001 EXTEND WT CNTL. Counseling pursuant to paragraph 6105 of reference (h) for Marines not making significant progress while on their first assignment or the extension to their first assignment to the BCP will occur after the initial 60-day probation period and, at a minimum, at the fourth month of the BCP assignment when the Marine is reevaluated by the MO and diagnosed as not having an underlying cause or disease. Subparagraph (2) provides the recommended standard language for this counseling.

(1) "Counseled this date concerning your extension on the Marine Corps BCP. You have made reasonable progress but have failed to reach your required body composition while on your first assignment to this Program. Therefore, you have been granted a one-time extension of 6 months to meet the body composition standards set forth in MCO P6100.12. You are advised that failure to take corrective action and meet your body composition standards or weight/body fat reduction goals, as prescribed by the medical officer, while on this 6 month extension will result in administrative separation from the USMC for either weight control failure or unsatisfactory performance per paragraphs 6206 and 6215 of MCO P1900.16E. Specific recommendations for corrective action are: Meet your weight/body fat reduction goals by losing the prescribed __ lbs and __ percent body fat per month for 6 months and actively participate in your diet and exercise program. Assistance is available through your chain of command. I was advised that within 5 working days after acknowledgment of this entry a written rebuttal could be submitted and that such a rebuttal will be filed on the document side of my SRB/OQR. I choose (to)(not to) make such a statement."

(2) "Counseled this date concerning your unsatisfactory performance while assigned to the Marine Corps BCP. Due to insufficient effort, you have not met your weight/body fat reduction goals as prescribed by the medical officer. You are advised that failure to take corrective action and reach your required body composition will result in administrative separation from the USMC for either weight control failure or unsatisfactory performance per paragraphs 6206 and 6215 of MCO P1900.16E. Specific recommendations for corrective action are: Meet your weight/body fat reduction goal and actively participate in your diet and exercise

program. Assistance is available through your chain of command. I am advised that within 5 working days after acknowledgment of this entry, a written rebuttal could be submitted and that such a rebuttal will be filed on the document side of my SRB/OQR. I choose (to)(not to) make such a statement."

>CH 1 c. Unlike the mandatory 6 month time period associated with the initial assignment to a BCP, Marines on an extension to a first assignment will be removed from the BCP when they return to HT/WT or BF standards; see paragraph 13 in this section for program removal instructions. A Marine not meeting standards at the end of the extension will be processed for administrative separation for failure to meet body composition standards per paragraph 6215 in MCO P1900.16F. No additional extensions can be granted.

>CH 1 11. Second Assignment. A Marine who has been assigned to and officially removed from the BCP at any time in their career, and fails to meet the weight and body fat standards a second time, may receive a one-time second assignment to the BCP (and RPCP). On a second assignment, Marines will be given 6 months with no extensions to conform to standards and will be counseled pursuant to paragraph 6105 of MCO P1900.16F with the recommended standard language. Unlike the mandatory 6 month time period associated with an initial assignment to a BCP, Marines on a second assignment to a BCP will be removed from the BCP when they return to HT/WT or BF standards. A Marine will be assigned to the Program as per paragraphs 2 through 10 of this section. A Marine successfully meeting the weight and body fat standards at the completion of the second assignment will be officially removed from the BCP (and RPCP) as stated in paragraph 13 in this Section. A Marine not meeting the standards when the second assignment expires will receive a 6105 counseling entry on the their SRB/OQR's Page 11 per paragraph 6206 or 6215 in MCO P1900.16F respectively, and will be processed for administrative separation. Extensions to a second assignment are not authorized.

> "Counseled this date concerning your reassignment to the Marine Corps BCP. Specifically, you failed to properly maintain your body composition standards as required by MCO P6100.12 for a second time. Your actions are unsatisfactory and not in accordance with the high state of readiness required by the USMC. You are advised that this subsequent assignment is for a 6-month period. Specific recommendations for corrective action are: Meet your weight/body fat reduction goals by losing the prescribed __ lbs and __ percent body fat per month and actively participate in your diet and exercise program. Assistance is available through your chain of command. You are advised that failure to take corrective action and reach the Marine Corps body composition standard will result in administrative separation from the USMC for either weight control failure or unsatisfactory performance per paragraphs 6206 and 6215 of MCO P1900.16E. Additionally, you are advised that MCO P6100.12 precludes assignment to the BPC for a third time in a Marine's career and that any subsequent failure to meet prescribed weight/body fat standards during your career will result in administrative separation. I am advised that within 5 working days after acknowledgment of this entry, a written rebuttal could be submitted and that such a rebuttal will be filed on the document side of my SRB/OQR. I choose (to)(not to) make such a statement."

12. Follow On Assignments. A Marine, assigned to the BCP on two separate occasions (e.g., first and second assignments), that fail to meet weight and body fat standards a third time will automatically be processed for administrative separation per paragraph 6215 of MCO P1900.16F. In such cases, the Marine will still be required to participate in the BCP and RPCP until discharged; no formal assignment is required. A prior counseling entry pursuant to paragraph 6105 of MCO P1900.16F is required to separate a Marine for either weight control failure or unsatisfactory performance per paragraphs 6206 and 6215 of MCO P1900.16F.

13. Program Removal. A Marine will not be removed from the BCP (and RPCP) assignment prior to the Program expiration date. A Marine will continue to participate in the BCP and RPCP as prescribed, regardless of progress, in order to ensure healthy weight and body fat loss techniques were employed and the weight and body fat loss is long term. A Marine successfully completing the assignment or extension will be administratively removed from the BCP (and RPCP).

a. Sixth Endorsement (appendix D, enclosure 7). The CO will complete the Sixth Endorsement, counsel the Marine on having successfully completed the BCP assignment, and have the Marine sign the endorsement. The Seventh Endorsement (appendix D, enclosure 8) only applies to Marines completing an extension to a first assignment.

b. The CO will direct the Admin Section to officially remove the Marine from the BCP, recording the following code/statement per reference (e) into MCTFS: TTC 339 000 FR WT CNTL. The Admin Section will record the Unit Diary number and date on the bottom of the endorsement and forward it to the Command PT Representative to be maintained in the Marine's BCP file.

c. Copies of completed BCP files will be kept on record and maintained by the Command PT Representative for 1 year.

14. Summary of Administrative Actions. Table 3-1 lists the administrative actions required for Marines who fail to meet the weight and body fat standards.

ADMINISTRATIVE ACTION	NUMBER OF ASSIGNMENTS	
	BCP 1st	BCP 2nd
Mandatory RPCP	YES	YES
Page 11 SRB/OQR Mandatory Entry	YES*	YES*
Adverse FitRpt or Substandard PRO/CON Marks	YES	YES
Eligible for Promotion	NO	NO
Eligible for PCS Transfer	YES	NO
Process for Administrative Separation	YES	YES
May Be Denied Re-enlistment	YES	YES
May Be Denied Special Schools	YES	YES
May Be Denied Recommendation for Promotion	YES	YES

Table 3-1.--Administrative Actions.

*A 6105 counseling entry will be recorded on the SRB/OQR's Page 11 the day the Marine is counseled on deficiencies, corrective actions to be taken, and assignment to the BCP. Additional Page 11 entries, per paragraph 6206 of MCO P1900.16F will only be required for a Marine who have not made satisfactory progress during the BCP assignment.

NOTE: A Marine who meets the Physical Performance Evaluation criteria will not be assigned to the BCP and will not receive an adverse FitRpt.

15. BCP and End of Active Service (EAS). CO discretion will be exercised when there are individual cases involving Marines who fail to meet weight and body fat standards just prior to their EAS.

a. Imminent EAS does not preclude a Marine who exceeds weight and body fat standards from being evaluated and processed for a BCP and RPCP assignment. A CO may evaluate and assign a Marine to participate in the BCP and RPCP up to the Marine's EAS date.

b. For a Marine who fails to meet weight and body fat standards upon completion of their 6-month BCP assignment or extension, and EAS is imminent, CO discretion will be exercised to decide whether or not measures to process the Marine for administrative separation is worth the administrative burden. CO may opt to extend the Marine's BCP and RPCP assignment up to the Marine's EAS date. A 6105 counseling entry on the SRB/OQR's Page 11 will be recorded, identifying the BCP failure, for Reserve component purposes.

16. Relocation Orders for Marines on the BCP

a. A Marine can be transferred, receive PCS or PCA orders during a first assignment, or (first assignment) extension to a BCP.

b. A Marine will not be transferred, but placed in an administrative hold status if he/she is currently participating in a second assignment to the BCP.

c. The Admin Section and Command PT Representative are required to ensure all Unit Diary entries, MO's Body Composition Evaluation(s), and BCP documents, are up to date and included in the Marine's BCP file, prior to being forwarded to the receiving command.

d. The command is required to forward by registered/receipt mail, the original BCP file to the command to which the Marine is being transferred, and keep a copy of the file on record for 1 year. This BCP file should be forwarded within 7 days of the Marine's official checkout date.

3202. Residential Obesity Program (ROP)

1. Marines that have been evaluated by a MO and determined to have an extreme weight and body fat increase due to an underlying or associated disease, will be recommended to participate in the ROP. The ROP is available to all Marines regardless of pay grade. A Marine will be sent to the ROP at the local ARC based on the following criteria:

a. A Marine has not previously participated in a ROP at a residential ARC during Marine Corps career.

b. A Marine has no medical history of having an eating disorder, (e.g., anorexia, bulimia, or other unresolved eating disorder).

c. A Marine must have a strong potential for continued military service supported by CO or related letters of recommendation.

d. A Marine must have at least 12 months time in service remaining on current military contract after the completion of the ROP for aftercare. [Waivers to remaining time in service criteria may be requested on a case-by-case basis from CMC (MMSR-4).]

2. Post-Residential Obesity Program (Rehabilitation). Upon completion of the ROP, the Marine will be assigned to the unit's RPCP for a minimum of 6 months. Weight and body fat reduction goals will vary depending upon the Marine's circumstances. Results must be monitored by the Command PT Representative and the MO assigned to the case. Entry in the ROP does not affect the BCP assignment process.

3. Rehabilitation Failure. In cases of rehabilitative failure, a Marine will be processed for administrative separation per paragraph 6206.1 of MCO P1900.16F. All rehabilitation failures must be reported to CMC (MMSR-4) via Naval message. Rehabilitation failure is based on the following circumstances:

a. Failure to demonstrate significant and consistent progress towards meeting Marine Corps weight and body fat standards during the 6-month rehabilitation period.

b. Continued increase in weight and/or body fat during the ROP or the aftercare period.

APPENDIX A

GLOSSARY

SECTION 1: MCO SHORT TITLES

MARCORMAN	THE MARINE CORPS MANUAL
MCO P5600.31G	MARINE CORPS PUBLICATIONS AND PRINTING REGULATIONS
MCO P1610.7E	PERFORMANCE EVALUATION SYSTEM (SHORT TITLE: PES)
MCO P1080.40C	MARINE CORPS TOTAL FORCE SYSTEM PERSONNEL REPORTING INSTRUCTIONS MANUAL (SHORT TITLE: MCTFSPRIM)
MCO P1070.12K	MARINE CORPS INDIVIDUAL RECORDS ADMINISTRATION MANUAL (SHORT TITLE: IRAM)
MCO P1900.16F	MARINE CORPS SEPARATION AND RETIREMENT MANUAL (SHORT TITLE: MARCORSEPMAN)
MCO P1080.20M	MARINE CORPS TOTAL FORCE SYSTEM CODES MANUAL (SHORT TITLE: MCTFSCODESMAN)
MCO P1040.31H	ENLISTED CAREER PLANNING AND RETENTION MANUAL
MCO P1400.31B	MARINE CORPS PROMOTION MANUAL, VOLUME 1, OFFICER PROMOTIONS (SHORT TITLE: MARCORPROMMAN, VOL 1, OFFPROM)
MCO P1400.32C	MARINE CORPS PROMOTION MANUAL, VOLUME 2, ENLISTED PROMOTIONS (SHORT TITLE: MARCORPROMMAN, VOL 2, ENLPROM)
MCO 5000.12D	MARINE CORPS POLICY ON PREGNANCY AND PARENTHOOD

SECTION 2: SHORT TITLES & ACRONYMS

APPROPRIATELY CREDENTIALED HEALTH CARE PROVIDER	ACHCP
ADMINISTRATIVE SEPARATION	ADSEP
ALCOHOL REHABILITATION CENTER	ARC
BODY COMPOSITION PROGRAM	BCP
CONVENIENCE OF THE GOVERNMENT	CofG
DISABILITY EVALUATION SYSTEM	DES
DIARY FEEDBACK REPORT	DFR
MARINE CORPS TOTAL FORCE SYSTEM	MCTFS
MEDICAL EVALUATION BOARD	MEB
MEDICAL OFFICER	MO
MILITARY TREATMENT FACILITY	MTF
NOT MEDICALLY QUALIFIED	NMED
PHYSICAL CONDITIONING PROGRAM	PCP
PHYSICAL EVALUATION BOARD	PEB
PERMANENT LIMITED DUTY	PLD
PARTIAL PHYSICAL FITNESS TEST	PPFT
PHYSICAL READINESS TRAINING	PRT
REQUIRED BUT DID NOT TAKE	RDNT
RESIDENTIAL OBESITY PROGRAM	ROP
REMEDIAL PHYSICAL CONDITIONING PROGRAM	RPCP
RETURN TO FULL DUTY	RTFD
TEMPORARY LIMITED DUTY	TLD

SECTION 3: DEFINITIONS

Appropriately Credentialed Health Care Provider (ACHCP). An individual who has specialized training or education in a health-related field. This may include administration, direct provision of patient care, or ancillary services. Such a professional is licensed, certified, or registered by a Government agency to provide specific health services in the field as an independent. An ACHCP, referred to throughout MCO P6100.12 as Medical Officer (MO), is a physician, physician's assistant, or nurse practitioner; Independent Corpsmen or regular Corpsmen are not considered ACHCPs.

Body Composition Program (BCP). Formerly known as the "Weight Control Program." Marines assigned to the BCP will receive assistance in reducing weight/body fat, in order to resume to a more healthy physical fitness state. The BCP assignment process and endorsements are contained in appendix D; the unit's administrative actions are contained in paragraph 3201 of MCO P6100.12.

Body Fat (BF). The amount of fat tissue a person has in ratio to lean body mass (or muscles, bones, organs, nervous tissue, and skin).

Diary Feedback Report (DFR). A report received in the Admin Section requiring action (e.g., DFR to notify the command that SNM's BCP assignment period has expired).

Light Duty. This is the status a MO assigns to a Marine when it is determined that a medical condition exists that interferes with the performance of duty, but the Marine is expected to return to full duty (RTFD). A MO may recommend up to 30 days light duty. When the light duty period expires, SNM will be medically reevaluated. If SNM is not RTFD status, then the Marine will be placed on medical hold and a Medical Evaluation Board (MEB) must be initiated and completed within the following 30 days.

Limited Duty. This is the status a MO assigns to a Marine when it is determined that a medical condition exists and will require an extended period of time before being RTFD. See Temporary Limited Duty and Permanent Limited Duty.

MEB. A MEB is a group of MOs at a local Military Treatment Facility (MTF) convened for the purpose of identifying the physical and mental qualifications of referred Marines to RTFD. The MEB is charged with determining whether a Marine's limitation(s) preclude their RTFD within a reasonable period of time or at all. A MEB will evaluate and report on the diagnosis, prognosis, possible RTFD, plan for further treatment, and make medical recommendation(s) for disposition.

MO. An ACHCP; see ACHCP.

Partial Physical Fitness Test (PPFT). Marines who are medically excused from performing a particular event of the PFT, as certified by a MO, are required to perform a PPFT. Marines will be required to obtain at least the minimum passing performance for the two events they are medically qualified, in order to pass the PPFT.

Permanent Limited Duty (PLD). PLD is a continuation on active duty in a permanent limited duty status, which is approved after a Marine has been determined "Unfit" by a PEB as a result of physical disability. PLD is assigned when a Marine is not expected to RTFD earlier than 16 months or it is projected that the condition cannot be corrected. A Marine can remain on active duty with a PLD status, provided criteria is met per MCO P1900.16F. PLD is requested through the PEB and approved by CMC (MMSR-4). PLD requests will be authorized by the PEB based on the best interest of the Marine Corps and the health of the Marine.

Physical Evaluation Board (PEB). The PEB acts on behalf of the SecNav to make determinations of fitness for continued military service, entitlement to benefits, disability ratings, and disposition of the medical case. A PEB is required to convene for cases that exceed the maximum 16 months of limited duty or when a Marine's initial diagnosis and prognosis do not anticipate a return to RFTD.

Physical Fitness Test (PFT). An instrument that measures the level of physical fitness of all Marines. It is a measurement of general fitness vice combat readiness and MOS capability. The standard PFT consists of three events that measure cardiovascular endurance, muscular strength and endurance, and mobility. Male Marines will perform dead-hang pullups, abdominal crunches, and a 3.0 mile run. Female Marines will complete the flexed arm hang, abdominal crunches, and a 3.0 mile run. These events are designed to test the strength and stamina of the upper body (shoulder girth), midsection, and lower body, as well as the efficiency of the cardiovascular system.

Physical Readiness Training (PRT). This training is comprised of general physical conditioning exercises that will develop and maintain strength, endurance, and the physical skills necessary to sustain individual Marines during combat.

Remedial Physical Conditioning Program (RPCP). Remedial physical conditioning is a process by which Marines are conditioned in a progressive manner to meet prescribed standards. The Program affords Marines the opportunity to rebuild themselves after a weight/body fat increase, injury/illness, pregnancy, or a period that lacked a structured fitness program.

Residential Obesity Program (ROP). Marines that have been evaluated by a MO and determined to have an extreme weight/body fat increase due to an underlying cause or associated disease, will be recommended to participate in the ROP.

Temporary Limited Duty (TLD). TLD is an assignment of limited duty for a specific period. Assignment to TLD can only be made after the completion of a MEB, and only if the prognosis is that the Marine can be RTFD within the specified period. The cumulative period of TLD will not exceed 16 months.

APPENDIX B

DAILY 16 PROGRAM

A. General

1. Purpose. The Daily 16 Program is a comprehensive series of warmup, conditioning, and cool-down exercises; it replaces the former Daily 7. The Daily 16 warmup exercises facilitate gradual distribution of blood flow to the muscles, preparing both the cardiovascular and musculoskeletal systems for the exercise session, by effectively targeting both the upper and lower body. The increased blood flow to the muscles produces a warming effect, increasing the elasticity of the muscles and connective tissues, which is believed to reduce injury risks. The Daily 16 conditioning exercises include traditional calisthenics that are safe in providing a total body workout, which can vary in duration, degree of difficulty, and level of intensity. The Daily 16 cool-down exercises (the same exercises used in the warmup) allow the body to gradually return to the pre-exercise state.

2. Scope. There are three different ways the Daily 16 Program can be used during physical training:

a. Daily 16 Warmup (D16WU). A series of warmup and dynamic stretching exercises that should be conducted prior to the main activity (e.g., formation run, obstacle course, circuit course, etc.), of every physical training session.

b. Daily 16 Workout (D16WO). A series of conditioning exercises that can be used as augmentation to another conditioning activity (circuit course, Physical Fitness Test (PFT), etc.), or can be used as a conditioning activity in of itself by simply increasing the number of repetitions or by slowing down the execution of the repetition.

c. Daily 16 Cool-Down (D16CD). A series of cool-down exercises that should be used as the final activity of a physical training session.

3. Daily 16 Warmup and Dynamic Stretching Descriptions

a. Warmup Exercises

(1) Heel to Toe Rocking. Starting position is standing with feet together and hands on hips. Rock back onto the heels, pause, then rock forward onto the toes and pause. Repeat 10 to 15 repetitions.

1

2

(2) Partial Squats. Starting position is standing with feet shoulder width apart, arms at the sides (1). Keeping the heels on the deck, partially squat until hands are near mid-calf. The knees should only bend to about 60 degrees, just short of a sitting position (2). Return to the starting position (3). Repeat 10 to 15 repetitions.

1/3

2

(3) Butt Kicks. Starting position is standing with feet shoulder width apart, hands on hips. Shift weight onto the right foot and quickly bend the left knee five times, bringing the left heel toward the buttocks. Switch legs and repeat. Repeat the whole cycle two or three times, until a total of 15 to 20 repetitions are done on each leg.

(4) Double Time in Place. Starting position is standing with arms at the sides. Slowly begin to run in place and gradually increase speed. While double-timing, conduct "punches to the front (1)" (throw easy punches to the front of the body). Switch to "punches to the sky (2)" (throw easy punches straight up to the sky). Finish with "arm circles (3)", (begin small then transition to large arm circles; repeat in other direction).

1 2 3

(5) Neck Flexion and Extension. Starting position is standing with feet shoulder width apart, hands on hips. Flex the neck forward bringing the chin toward the chest, pause. Extend the head back and pause. Repeat for 5 to 10 repetitions.

(6) Neck Lateral Flexion. Starting position is standing with feet shoulder width apart, hands on hips. Tilt the head to the left side, bringing the left ear toward the left shoulder, pause. Switch sides and repeat. Repeat for 5 to 10 repetitions.

(7) Trunk Flexion and Extension. Starting position is standing with feet shoulder width apart, hands on hips (1). Extend the trunk backward bringing a slight hyperextension of the trunk and pause (2). Repeat for 5 to 10 repetitions.

1

2

(8) Trunk Lateral Flexion. Starting position is standing with feet shoulder width apart, hands on hips. Bend the trunk to the left side (1), pause, then switch to the right side (2) and pause. Repeat for 5 to 10 repetitions. Proceed to the Stretching Exercises.

1

2

b. Dynamic Stretching Exercises

(1) Upper Back Stretch. Starting position is standing with feet shoulder width apart. Extend the arms and clasp the hands in front of the chest. Push the arms forward rounding the shoulders and upper back. Hold the position for 10 seconds and breath naturally. The stretch should be felt over the upper back.

(2) Chest Stretch. Starting position is standing with feet shoulder width apart. Clasp hands together behind the lower back, palms up. Pull the arms up toward the head. Hold the position for 10 seconds and breath naturally. The stretch should be felt in the front of the chest and shoulders.

(3) Modified Hurdler Stretch. Starting position is in the sitting position. Extend the left leg out while tucking the right leg in front of the hips with the knee pointing outward. Bend the torso forward toward the left knee. The stretching should be felt over the back of the left thigh. Hold the position for 10 seconds and breath naturally. Switch sides and repeat.

(4) Hip and Back Stretch. Starting position is in the sitting position. Extend the right leg straight out and cross the left leg over the right leg by bending the left knee and placing the left foot on the deck next to the right knee. Turn the upper torso to the left pushing the left knee to the right with the right elbow. Hold the position for 10 seconds and breath naturally. The stretch should be felt over the lower back and left hip. Switch sides and repeat.

(5) Groin Stretch. Starting position is in the sitting position with both knees bent and the bottoms of the feet together. Grasp the feet and gently push the knees with the elbows toward the deck. Hold the position for 10 seconds and breath naturally. The stretch should be felt over the inside of both thighs.

(6) Calf Stretch. Starting position is standing with arms at the sides. Place the left foot approximately 2 feet forward and slightly bend the right knee. Lean forward toward the left foot pointing the left toes up to the sky. Hold the position for 10 seconds and breath naturally. Grabbing the left foot and gently pulling it towards you can increase the level of intensity. The stretch should be felt over the left calf. Switch sides and repeat.

(7) Iliotibial Band (ITB) Stretch. Starting position is standing with arms at the sides. Place the left foot behind and a few inches to the right of the right foot. Bring the left arm over the head. Place your body weight on the left leg and bend at the waist to the right. Hold the position for 10 seconds and breath naturally. The stretch should be felt over the left hip. Switch sides and repeat.

(8) Hamstring Stretch. Starting position is lying down with the back flat against the deck. Bring the left knee toward the chest grasping the left leg just below the knee. Gently straighten the left knee and hold for the count. The right leg should remain on the deck. Hold the position for 10 seconds and breath naturally. The stretch should be felt on the back of the left thigh. Switch sides and repeat.

(9) Neck Stretch. Starting position is standing with feet shoulder width apart. Move both arms behind the back and grasp the left wrist with the right hand. Tilt the head to the right and pull the left arm to the right. Hold the position for 10 seconds and breath naturally. The stretch should be felt over the left shoulder and left side of the neck. Switch sides and repeat.

(10) Triceps Stretch. Starting position is standing, arms at the sides. Bend the left elbow and bring the left arm up and back placing the left hand between the shoulder blades. Gently pull the left elbow with the right hand behind the head. Hold the position for 10 seconds and breath naturally. The stretch should be felt over the back of the upper arm. Switch sides and repeat.

(11) Quadriceps Stretch. Starting position is lying down on the left side. Bend the left hip and knee to 90 degrees. Grasp the right ankle with the right hand and pull the right knee straight back. Do not hyperextend the lower back. Hold the position for 10 seconds and breath naturally. The stretch should be felt over the front of the right thigh. Switch sides and repeat.

(12) Lying Down ITB Stretch. Starting position is lying down on the deck. Bring the left leg with the knee straight across the body (1). Hold the position for 10 seconds and breath naturally. The stretch should be felt over the left hip. Switch sides and repeat (2).

1

2

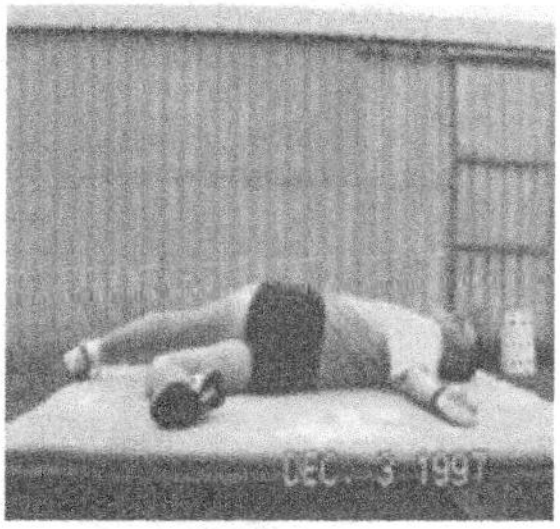

(13) Posterior Shoulder Stretch. Starting position is standing with feet shoulder width apart, arms at the sides. Bend the left elbow and bring the left arm across the chest. Give a gentle pull with the right hand. Hold the position for 10 seconds and breath naturally. The stretch should be felt over the posterior left shoulder. Switch sides and repeat.

(14) Hip Flexor Stretch. Starting position is standing, hands on hips. Step the left foot forward 3 to 4 feet. Place the right knee on the deck. Gently move the left knee forward. Hold the position for 10 seconds and breath naturally. The stretch should be felt over the front of the right thigh and hip. Switch sides and repeat.

(15) Single-Leg Lower Back Stretch. Starting position is lying with the back flat against the deck. Bring the right knee toward the chest grasping the right knee. Gently pull the knee tight into the chest. The left leg should remain on the deck. Hold the position for 10 seconds and breath naturally. The stretch should be felt along the lower back to the right buttock. Switch sides and repeat.

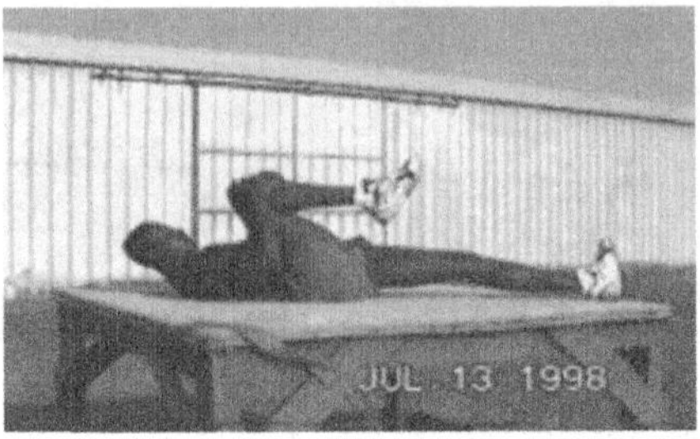

(16) Prone Abdominal Stretch. Starting position is lying on the stomach with the hands placed near the shoulders as if in the down position of a pushup. Slowly raise the upper body, keeping the waist on the deck. Hold the position for 10 seconds and breath naturally. The stretch should be felt over the abdomen.

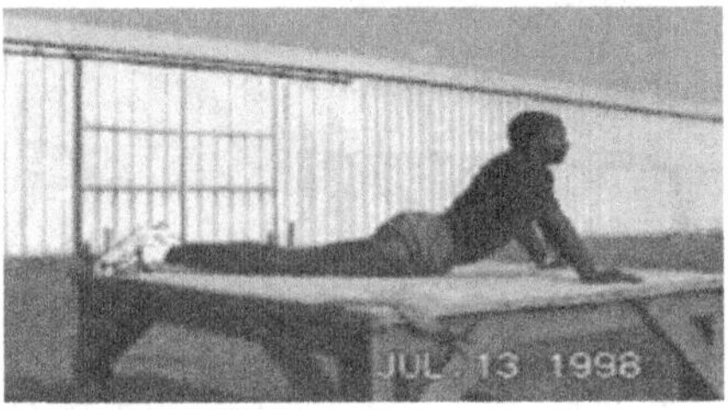

4. Daily 16 Conditioning Exercise Descriptions

a. Pushups. Starting position is lying on your stomach with hands shoulder width apart, toes on the deck, and elbows, back, and knees straight. On the first and third count, lower the chest to the deck; bend the elbows to at least 90 degrees (1/3). On the second and fourth count, extend the arms back to the starting position (2/4). This exercise conditions the chest, primarily in the anterior shoulder region, and secondarily, the triceps.

1/3

2/4

b. Crunches. Starting position is lying on the back with the hips bent to 90 degrees and the knees bent, feet off the deck. Bend the elbows to 90 degrees and fold across the chest or rib cage. On the first and third count raise the upper torso off the deck touching the thighs with the forearms (1/3). On the second and fourth count, return to the starting position (2/4). The arms must remain in constant contact with the chest/rib cage when executing the crunch. This should be done in a slow and controlled manner. This exercise conditions the abdominal muscles.

1/3

2/4

c. Dirty Dogs. Starting position is on the hands and knees. On the first and third count raise the left leg to the side, while keeping the knee bent (1/3). On the second and fourth count return the leg to the starting position (2/4). Switch sides and repeat. This exercise conditions the hip abductors.

1/3

2/4

d. Wide Pushups. Starting position is lying on the stomach with hands beyond shoulder width apart, toes on the ground, and elbows, back, and knees straight. On the first and third count, lower the chest to the deck, bending the elbows to at least 90 degrees (1/3). On the second and fourth count, extend the arms back to the starting position (2/4). This exercise conditions the chest, primarily in the anterior shoulder region, and secondarily, the triceps. With the wider hand position, the chest muscles increase their workload.

1/3

2/4

e. Dive Bomber Pushups. Starting position is lying on the stomach with hands and toes on the deck, and elbows and knees straight. The hands will be slightly beyond shoulder width apart. The hips will be raised up and the shoulders will be behind the hands. On the first count, lower the chest down and forward to the deck, the shoulders will be even with the hands (1). On the second count, continue forward extending the elbows where now the shoulders are in front of the hands (2). On the third count, reverse the direction lowering the chest down and back to the deck, the shoulders will be even with hands (3). On the fourth count, continue back and up to the starting position (4). This exercise is done in a smooth, continuous motion.

This exercise conditions the chest and anterior shoulder primarily, and secondarily, the triceps through a greater ranger of motion.

1

2

3

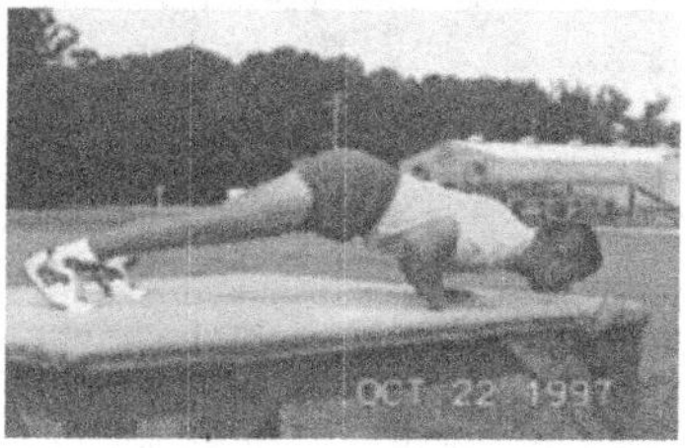

4

f. Elbow to Knee Crunches. Starting position is lying on the back with the right foot flat on the deck and, the left foot crossed over the right knee, and the arms crossed over the chest. On the first and third count, raise the upper torso off the deck rotating to the left touching the right elbow to the left thigh (1/3). On the second and fourth count, return to the starting position (2/4). This exercise should be done in a slow and controlled manner. Switch sides and repeat. This exercise conditions the abdominal muscles with more emphasis on the oblique.

1/3

2/4

g. <u>Side Crunches</u>. Starting position is lying on the left side with the left arm across the chest and right arm along the side of the body. On the first and third count, raise the upper torso and feet off the deck sliding the right hand down the thigh (1/3). On the second and fourth count, return to the starting position (2/4). Switch sides and repeat. This exercise conditions the abdominal muscles with emphasis on the internal oblique and external oblique.

1/3

2/4

h. <u>Prone Flutter Kicks</u>. Starting position is lying on the stomach. On the first count raise the left leg off the deck while the right leg remains on the deck (1). On the second count, lower the left leg to the starting position (2). On the third count, raise the right leg off the deck while the left leg remains on the deck (3). On the fourth count, lower the right leg to the starting position (4). This exercise conditions the muscles that extend the hip and back.

1

2

3

4

i. Back Extension. Starting position is lying on the stomach with hands behind the head. On the first and third count, raise the upper torso and legs off the deck (1/3). On the second and fourth count, lower the upper torso and legs to the starting position (2/4). This exercise conditions the muscles that extend the back.

1/3

2/4

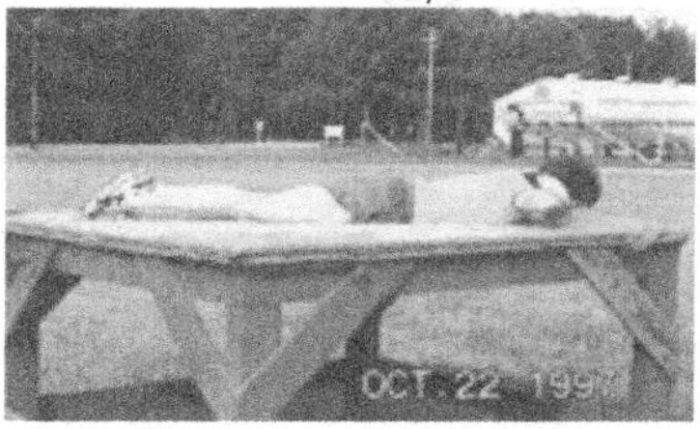

j. Donkey Kicks. Starting position is on the hands and knees. On the first and third count, kick the left leg back and up straightening the knee (1/3). On the second and fourth count, bend the knee and hip bringing the left knee into the chest (2/4). The back should not hyperextend during this exercise. Switch sides and repeat. This exercise conditions the muscles that extend the hip.

1/3

2/4

k. Hip Adduction. Staring position is lying on the left side with the right leg bent setting the right foot in front of the left knee. On the first and third count, raise the straight left leg off the deck squeezing the thighs together (1/3). On the second and fourth count, lower the left leg to the starting position (2/4). The left toes should be pointing straight-forward not to the sky. Switch side and repeat. This exercise conditions the muscles on the inner thigh.

1/3

2/4

1. <u>Side Leg Raises</u>. Starting position is lying on the left side with the left knee bent, the hip vertical and the toes on the right foot pointing forward not to the sky. On the first and third count, raise the right leg approximately 18 inches leading with the heel (1/3). The toes will still point forward, not to the sky. On the second and fourth count, lower the right leg to the starting position (2/4). Switch sides and repeat. This exercise conditions the muscles on the side of the hip and thigh.

1/3

2/4

m. Steam Engines. Starting position is standing with the feet shoulder width apart and hands behind the head. On the first count, touch the right elbow to the left knee by bending and raising the left knee and twisting and bending the upper torso to the left (1). On the second count, return to the starting position (2). On the third count, touch the left elbow to the right knee by bending and raising the right knee and twisting and bending the upper torso to the right (3). On the fourth count, return to the starting position (4). This exercise should be done in a slow and controlled cadence. This exercise conditions the abdominal muscles, primarily the oblique, and secondarily, the hip and thigh.

1

2

3

4

n. Lunges. Staring position is standing with the feet shoulder width apart and hands on the hips. On the first count, touch the right knee to the deck by stepping forward with the left foot and bending both knees (1). On the second count, return to the starting position (2). On the third count, touch the left knee to the deck by stepping forward with the right foot and bending both knees (3). On the fourth count, return to the starting position (4). Do not bend the forward knee more than 90 degrees. This exercise should be done in a slow and controlled cadence. This exercise conditions the muscles that extend the hip and knee of the forward leg.

1

2

3

4

o. <u>Side Straddle Hops</u>. Starting position is standing with the feet together and arms at the sides. On the first and third count, jump and land with both feet just beyond shoulder width apart while bringing the hands together overhead (1/3). On the second and fourth count, jump back to the starting position (2/4). This exercise is done in a smooth, continuous motion. This exercise conditions the upper arm, hip, and thigh.

1/3

2/4

5. Daily 16 Warmup/Cool-Down, Stretching and Exercise Cards

STRETCHING CARD A

Chest Stretch
Triceps Stretch
Posterior Shoulder Stretch
Iliotibial Band (ITB) Stretch
Modified Hurdler Stretch
Hip and Back Stretch
Quadriceps Stretch
Lower Back Stretch
Abdominal Stretch

EXERCISE CARD 1

Wide Pushups
Donkey Kicks
Crunches
Dive Bomber Pushups
Dirty Dogs
Side Crunches
Back Extensions
Lunges
Side Straddle Hops

STRETCHING CARD B

Upper Back Stretch
Chest Stretch
Shoulder & Neck Stretch
Triceps Stretch
Posterior Shoulder Stretch
Lying Down (ITB) Stretch
Modified Hurdler Stretch
Groin Stretch
Calf Stretch

EXERCISE CARD 2

Pushups
Crunches
Side Leg Raises
Diamond Pushups
Elbow-to-Knee Crunches
Prone Flutter Kicks
Hip Abduction
Lunges
Steam Engines

STRETCHING CARD C

Active Hamstring Stretch
Groin Stretch
Hip Flexor Stretch
Lower Back Stretch
ITB Stretch
Calf Stretch
Neck Stretch
Upper Back Stretch
Chest Stretch

WARM-UP/COOL-DOWN CARD

Toe-Heel Rocking
Partial Squats
Butt Kicks
Trunk Bends
Neck Bends
Run in Place
Punch to the Front
Punch to the Sky
Arm Circles
*Stretch Card (10 Seconds)
*Exercise Cards

APPENDIX C

REMEDIAL PHYSICAL CONDITIONING PROGRAM (RPCP)

1. Purpose. Remedial physical conditioning is a process by which Marines are conditioned in a progressive manner to meet prescribed physical fitness and body composition standards. The RPCP goal is to provide challenging conditioning sessions, using the spectrum of aerobic conditioning, resistance conditioning, and other related exercises that will afford the Marines an opportunity to rebuild themselves after a weight or body fat increase, injury/illness, pregnancy, or a period that lacked a structured fitness program.

2. Assignment to RPCP. Participation in the RPCP is a mandatory requirement for Marines who fail the Physical Fitness Test (PFT) or who are assigned to the Body Composition Program (BCP). Marines will also be assigned to the RPCP when they consistently display problems in MOS-specific tasks, (e.g., conditioning marches, etc.) or struggle to meet the minimum standard for an event during a semi-annual PFT. Performance on unit formation runs will not be used as RPCP entry/exit criteria. Marines recovering from injuries or prolonged illness may be assigned to the RPCP if a formal remedial program would be beneficial. In the later case, COs must take into account any followup medical care (therapy) requirement and tailor the RPCP to accommodate those physical limitations or restrictions imposed by the Medical Officer (MO). Deficiencies, which can be improved or corrected by a command-directed remedial conditioning program, fall into several categories:

a. Lack of strength in one or more body parts;

b. Lack of overall cardiovascular or muscular endurance;

c. Lack of adherence to body composition standards;

d. Lack of mobility (posture, balance, stability, agility, coordination, power, speed and flexibility);

e. Lack of motivation and dedication; and/or

f. Lack of education and experience in health and fitness.

3. Organization of the RPCP

a. Leadership. The intent of the RPCP is not punitive in nature, but designed to encourage improvement in overall fitness and body composition. Unit leaders and Command Physical Training (PT) Representatives should assist Marines in clearly identifying their deficiencies, developing specific strategies to improve, and tracking progress. The RPCP should focus on realistic goal setting, healthy life-styling strategies, progressive PT, dietary measures, and testing and evaluation. The RPCP should be challenging, employ a cross-training philosophy, and ensure safety measures are taken to avoid over-training and potential overuse injuries.

b. Use of Resources. This Appendix will provide a few examples of remedial conditioning programs units can employ or use as a guide for their own program designs. Command PT Representatives can also access the resources available at Semper Fit Fitness Centers, which are staffed by qualified professionals and offer a variety of health and fitness services.

c. Organization. Remedial conditioning should be conducted in small groups, (e.g., no larger than 10-15 Marines per group), thereby allowing for more focused instruction, oversight, and reinforcement. It is recommended that the RPCP initially be organized by grouping Marines by area of deficiency and level thereof. Prescribed exercise loads should be commensurate with their abilities, and whenever possible, linking a strong Marine with a weaker Marine. An organization as such will support performance and improvement.

(1) As the RPCP and individual fitness and performance levels progress, Marines should be reorganized to concentrate on remaining deficiencies, or combined to conduct specific PT events, such as conditioning marches or team sports. For example, a group that exhibits a deficiency in upper body strength would execute a program that includes pullups, rope climbing, rifle PT, log drills, and weight training. A second group may exhibit deficiencies in cardiovascular respiratory endurance, and therefore would execute a program that includes individual and ability group running, grass drills, circuit training, and fartlek running.

d. Measurement. Continual measurement of a Marine's deficient area(s) is the only way to gauge progress. While testing should not take the place of training sessions geared towards improving performance, periodic testing (e.g., every 2 weeks) is an important step of any remedial conditioning program. Testing, PFT, and body composition evaluations for example, should commence at the end of the first week of a RPCP.

4. RPCP

a. Circuit Training and Exercise Drills. Circuit training is described as a specific training routine in which a variety of fitness discrepancies are addressed. It can accommodate a large number of Marines with little supervision, provide variety and challenge with its progressive programming, and allow participants to progress at their own rate. The goal of circuit training is to develop strength and endurance through a systematic and progressive conditioning program that involves stations where specific exercises are performed. These exercises are performed vigorously for a short period of time before moving to a follow-on station. The vigorous activity in short periods of time provide a near maximal quality training session, assuring progressive overload in an organized manner. The exercise selected for each station and the arrangement of the stations is determined by the objective of the circuit.

(1) Free Circuit. Free (portable) circuits use equipment that can be readily transported or set up for use regardless of the training area and environment. A simplified course can be executed on a basketball court, a field, or on the deck of a ship. In a free circuit, there is no set time for staying at each station and no signal to move from one station to the next. Marines work at their own pace, doing a fixed number of repetitions (rep) at each station. Progress is measured by the time needed to complete a circuit. However, in remedial training, the quality and number of repetitions should be monitored.

(2) Fixed Circuit. Fixed circuits require equipment that is normally in place and can be used on a regular basis. In a fixed circuit, a specific length of time is set for each station. The time is monitored with a stopwatch, with Marines rotating through the stations on command. To increase the intensity or difficulty of a fixed circuit:

(a) maintain the time for completion, but increase the number of reps;

(b) increase the time per station along with the number of reps; and,

(c) increase the number of times a circuit is negotiated.

b. Number of Stations/Circuit Training Reps. The goal of the circuit, as well as the time and equipment available will determine the number of stations. A circuit geared for the objective of developing upper or lower body strength may need as little as six to eight stations, while a cardiovascular strength and endurance circuit may have as many as 20 stations. The number of Marines assigned to negotiate a single circuit station is determined by the total number of participants, the number of stations, and the level of supervision available. To achieve the desired training effect, the circuit may have to be repeated several times. For example, a 10-station circuit requiring Marines to exercise for 30 seconds at each station and 15 seconds to move between stations may result in Marines completing the entire evolution three times. The entire training session, designed to take between 45-60 minutes, includes a warmup and cool-down period (e.g., Daily 16). As the conditioning level increases, consideration can be given to either increase exercise periods from 30 to 45 and 60 second intervals, or the addition of a fourth rotation through the circuit. Stations should be arranged in a sequence that allows for adequate recovery time between high/low intensity exercises or by alternating different muscles groups.

(1) Cardiovascular Endurance Circuit #1. This circuit is designed to improve the cardio-respiratory system by exercising hard at each of the eight exercise stations, with a steady recovery time (slow jog) between stations. Recommended course length is approximately 1/4 to 1/2 mile in length.

Station #1:	Ab Crunch	20 reps
Station #2:	Bend and Thrust	20 reps
Station #3:	Sprint	1 rep
Station #4:	Inclined Pushups	20 reps
Station #5:	Star Jumps	20 reps
Station #6:	Dorsal Raise	20 reps
Station #7:	Knee Lifts	20 reps
Station #8:	Dips	20 reps

(2) Cardiovascular Endurance Circuit #2. Recommended course length is 1/2 to 1.0-mile, with 14 events/stations:

Station #1:	Stationary Run	30 seconds
Station #2:	Pushup	30 seconds
Station #3:	Side Straddle Hop	30 seconds
Station #4:	Crunches	30 seconds
Station #5:	Ski Jumps	30 seconds
Station #6:	Flutter Kicks	30 seconds
Station #7:	Bend and Reach (slow reps)	30 seconds
Station #8:	Wide Pushups	30 seconds
Station #9:	Bicycle	30 seconds
Station #10:	Knee Bender	30 seconds
Station #11:	Steam Engines (Daily 16)	30 seconds
Station #12:	Twisting Situp	30 seconds
Station #13:	Lunges (Daily 16)	30 seconds
Station #14:	All-Fours Run	30 seconds

(3) Upper Body Strength Circuit #2. This circuit is designed to improve upper body strength and endurance, and is normally conducted following a cardiovascular training session. Proper form and full range of motion are important. Eight stations are set up over a course length of approximately 500 yards.

Station #1:	Crunches	20 reps
Station #2:	Pullups	3 sets @ max effort
Station #3:	Dorsal Raises	20 reps
Station #4:	Tricep Dips	40 reps
Station #5:	Situps	2 sets of 5 reps
Station #6:	Rope Heaves	20 reps
Station #7:	Pushups	20 reps
Station #8:	Sprint	1 rep

(4) Upper Body/Abdominal Strength Circuit #3. This circuit is designed to encompass a cardiovascular/anaerobic (sprint) portion that incorporates eight exercise stations with no set course length required. This circuit is conducted without a set time signal, e.g., Marines move immediately to a station after each exercise is completed. Recommended one to two complete course rotations, with a possible time decrease to 20 seconds (vice 30) on the second running.

Station #1:	Elevated Pushup	30 seconds
Station #2:	Twisting Situp	30 seconds
Station #3:	Parallel Dips	30 seconds
Station #4:	Ab Crunch (feet unsecured)	30 seconds
Station #5:	Wide Pushup	30 seconds
Station #6:	Flutter Kicks	30 seconds
Station #7:	Diamond Pushup	30 seconds
Station #8:	Bicycle	30 seconds

(5) Lower Body/Upper Body Strength (Weight Room) Circuit #4. This circuit is designed to follow a cardiovascular session. There are 13 exercise stations and the circuit should be conducted in one to two complete rotations. Each exercise should encompass slow and controlled movements, striving for muscle failure within 8-12 repetitions. A preliminary session, for Marines to become familiar with weight room layout and starting weight limits and execution is recommended.

Station #1:	Leg Press	8-12 reps
Station #2:	Leg Raise	8-12 reps
Station #3:	Leg Extension	8-12 reps
Station #4:	Leg Curl	8-12 reps
Station #5:	Heel Raise	8-12 reps
Station #6:	Bench Press	8-12 reps
Station #7:	Seated Row	8-12 reps
Station #8:	Military Press	8-12 reps
Station #9:	Lat Pulldown	8-12 reps
Station #10:	Shrugs	8-12 reps
Station #11:	Triceps Extension	8-12 reps
Station #12:	Biceps Curl	8-12 reps
Station #13:	Incline Situp	8-12 reps

c. Fartlek Training. Fartlek training is designed to develop strength, stamina, and endurance by combining running (cardiovascular) and calisthenics (strength and endurance). Fartleks consist of various running exercises, which strive to overload the cardiovascular system by allowing only partial recovery between each intense training segments; this sequence is repeated for the duration of the training session. Fartleks are normally conducted over a minimum 3-mile course, preferably with varying terrain and intermittent exercise stations to create the work overload.

(1) Initial/Baseline Fartlek:

- Warmup (Daily 16 warmup routine)
- Warmup run at easy pace for 5-10 minutes
- Run fast, steady pace for 3/4-1 3/4 miles
- Walk/Jog at moderate pace for 5 minutes (recovery)
- Run easily - sprinting 15-20 meters occasionally
- Run full speed uphill for 175-200 meters
- Cool-down by running easily for 1/2-1 mile
- Cool-down (Daily 16 cool-down routine)

(2) Cardiovascular/Muscular Endurance Fartlek: Recommended course length is 3.0-3.5 miles. Location and spacing of exercise stations is normally based on terrain. After a unit orientation run, the Fartlek course should be run as an individual effort.

Station #1:	Mountain Climbers	20 reps
Station #2:	Bends and Thrust	20 reps
Station #3:	Ab Crunches (feet unsecured)	20 reps
Station #4:	Sprint	175-200 meters
Station #5:	Situp w/Twist	20 reps
Station #6:	Pullups or Pushups	10-20 reps
Station #7:	Dorsal Raises	20 reps
Station #8:	Star Jumps	20 reps
Station #9:	Squat Thrusts	20 reps

5. RPCP Schedules

a. Upper Body Strength (Weekly) Circuit. A 3 day per week routine is recommended to improve overall upper body strength deficiencies. Conduct tests every 2 weeks and increase weights and reps as necessary.

Monday: Weight Training

Pullups	2 sets to fatigue
Lat Pulldown*	2 sets of 15 reps
Seated Row*	2 sets of 15 reps
Biceps Curl (EZ-Curl Bar)*	2 sets of 15 reps
Alternating Bicep Curls (Dumbbell)	2 sets of 15 reps

* Strive for momentary muscular fatigue at the end of each set. Allow 45-60 seconds rest between sets and 2-3 minutes rest between exercises.

Tuesday: Cross Training - Cardiovascular/Abdominal Training

Wednesday: Pullup/Pushup/Crunch Pyramid **

Set #1:	1 Pullups/2 Pushups
Set #2:	2 Pullups/4 Pushups
Set #3:	3 Pullups/6 Pushups/25 Crunches
Set #4:	4 Pullups/8 Pushups
Set #5:	5 Pullups/10 Pushups/25 Crunches
Set #6:	4 Pullups/8 Pushups
Set #7:	3 Pullups/6 Pushups/25 Crunches
Set #8:	2 Pullups/4 Pushups
Set #9:	1 Pullups/2 Pushups

** Decrease/Increase reps in pyramid based on the Marine's ability and progression of strength, the minimal rest between sets, and spotter assistance necessary to complete required reps.

Thursday: Rest

Friday: Repeat Monday training session

b. Strength and Endurance Training Schedule (20 weeks). This training schedule's primary fitness element is running, but also strives to combine aspects of strength, endurance, and flexibility. The running portion of the program provides the base of cardiovascular endurance necessary to augment all other physical skills. In addition to the progressive overload running program, strength training and flexibility exercises (e.g., Daily 16), addressing lower and upper body resistance training must be incorporated. Weight training can augment or be substituted for the Daily 16 exercises. Each training period will include a warmup prior to exercise and a cool-down session to follow.

WEEK 1,2	Endurance	Strength
Monday	Run 40 min @ 10 min/mile pace	Lower Body Resistance Daily 16 or Weight Training (1 set)
Tuesday	None	Upper Body Resistance - Daily 16 or Weight Training (1 set)
Wednesday	Run 30 min @ 10 min/mile pace	Lower Body - (1 set)
Thursday	None	Upper Body - (1 set)
Friday	Run 30 min @ 10 min/mile pace	Lower Body - (1 set)

WEEK 3,4	Endurance	Strength
Monday	Run 44 min @ 10 min/mile	Lower Body - (1 set)
Tuesday	None	Upper Body - (1 set)
Wednesday	Run 33 min @ 10 min/mile	Lower Body - (1 set)
Thursday	None	Upper Body - (1 set)
Friday	Run 33 min @ 10 min/mile	Lower Body - (1 set)

WEEK 5,6	Endurance	Strength
Monday	Run 48 min @ 10 min/mile	Lower Body - (2 sets)
Tuesday	None	Upper Body - (2 sets)
Wednesday	Run 36 min @ 10 min/mile	Lower Body - (2 sets)
Thursday	None	Upper Body - (2 sets)
Friday	Run 36 min @ 10 min/mile	Lower Body - (2 sets)

WEEK 7,8	Endurance	Strength
Monday	Run 53 min @ 10 min/mile	Lower Body - (2 sets)
Tuesday	None	Upper Body - (2 sets)
Wednesday	Run 40 min @ 10 min/mile	Lower Body - (2 sets)
Thursday	None	Upper Body - (2 sets)
Friday	Run 40 min @ 10 min/mile	Lower Body - (2 sets)

WEEK 9	Endurance	Strength
Monday	Run 53 min @ 9 min/mile	Upper Body - (3 sets)
Tuesday	None	Lower Body - (3 sets)
Wednesday	Run 40 min @ 10 min/mile	Upper Body - (2 sets)
Thursday	None	Lower Body - (3 sets)
Friday	Run 40 min @ 10 min/mile	Lower Body - (2 sets) Power - 2 reps

WEEK 10	Endurance	Strength
Monday	Run 53 min @ 8.5 min/mile	Upper Body - (2 sets)
Tuesday	Intervals: 4x 400m sprints in 2 min	Lower Body - (3 sets)
Wednesday	Run 40 min @ 8.5 min/mile	Upper Body - (2 sets)
Thursday	None	Lower Body - (3 sets) Power - 3 reps
Friday	Run 40 min @ 8.5 min/mile	Upper Body - (2 sets)

WEEK 11	Endurance	Strength
Monday	Run 53 min @ 8.5 min/mile	Upper Body - (3 sets)
Tuesday	Intervals: 4x 400 m sprints in 2 min	Lower Body - (2 sets)
Wednesday	Run 40 min @ 8.5 min/mile	Upper Body - (3 sets)
Thursday	None	Lower Body - (2 sets) Power - 4 reps
Friday	Run 40 min @ 8.5 min/mile	Upper Body - (3 sets)

WEEK 12	Endurance	Strength
Monday	Run 58 min @ 8.5 min/mile	None
Tuesday	Intervals: 5x 400 m sprints in 2 min (4 min rest)	Upper/Lower Body - (2 Sets)
Wednesday	Run 40 min @ 8.5 min/mile	None
Thursday	None	Upper/Lower Body - (3 sets) Power - 4 reps
Friday	Run 30 min @ 7.5 min/mile	None

WEEK 13	Endurance	Strength
Monday	Run 58 min @ 8.5 min/mile	None
Tuesday	Intervals: 6x 400 m sprints in 2 min (4 min rest)	Upper/Lower Body - (3 sets)
Wednesday	Run 40 min @ 8.0 min/mile	None
Thursday	None	Upper/Lower Body - (3 sets) Power - 4 reps
Friday	Run 30 min @ 7.0 min/mile	None

WEEK 14	Endurance	Strength
Monday	Run 58 min @ 8.5 min/mile	None
Tuesday	Intervals: 6x 400 m sprints in 2 min (3:30 min rest)	Upper/Lower Body - (3 sets)
Wednesday	Run 40 min @ 8.0 min/mile	None
Thursday	None	Upper/Lower Body - (3 sets) Power - 4 reps
Friday	Run 30 min @ 7.0 min/mile	None

WEEK 15	Endurance	Strength
Monday	Run 30 min (easy pace)	Upper/Lower - (1 set)
Tuesday	None	None
Wednesday	Run 20 min (easy pace)	None
Thursday	None	None
Friday	Run 20 min (easy pace)	Upper/Lower - (1 set)

WEEK 16	Endurance	Strength
Monday	Run 30 min (easy pace)	Upper/Lower - (1 set)
Tuesday	None	None
Wednesday	Run 20 min (easy pace)	None
Thursday	None	None
Friday	Run 20 min (easy pace)	Upper/Lower - (1 set)

WEEK 17	Endurance	Strength
Monday	Run 30 min (easy pace)	None
Tuesday	None	Lower Body - (2 sets) Power - 2 sets
Wednesday	Run 20 min (easy pace)	None
Thursday	None	Lower Body - (2 sets) Power - 2 sets or Obstacle Course
Friday	Run 15 min (easy pace)	None

WEEK 18	Endurance	Strength
Monday	Team Sport	None
Tuesday	None	Lower Body - (1 set) Power - 1 set
Wednesday	Team Sport	None
Thursday	None	Lower Body - (1 set) Power - 1 set or Obstacle Course
Friday	Team Sport	None

WEEK 19	Endurance	Strength
Monday	Team Sport	None
Tuesday	None	None
Wednesday	Team Sport	None
Thursday	None	None
Friday	Team Sport	Obstacle Course

WEEK 20	Endurance	Strength
Monday	None	Upper/Lower - (2 sets)
Tuesday	Team Sports	None
Wednesday	None	Upper/Lower - (2 sets)
Thursday	Team Sports	Obstacle Course
Friday	None	Upper/Lower - (2 sets)

On week 21, begin the training cycle again, starting at week 1.

The 20-week PT cycle uses a modified form of periodization training. No athlete can maintain a continuous competitive edge - any attempt to do so will result in an overtrained state where mental and physical abilities will deteriorate.

Interruptions in the training schedule are expected. As a general guideline, for each week during which scheduled PT is not done, the unit should go back 2 weeks from the last week to safely restart the program.

c. <u>5-Day/6-Week Cardio/Muscular Strength Conditioning Program</u>

MONDAY	TUESDAY	WEDNESDAY	THURSDAY	FRIDAY
WEEK 1				
PFT	Resistance Training (RT)	Aerobic Training (AT) Ability Group Run (ABGR) 3-mi Upper Body Strength Training (UBS)	RT	Hike: 5-mi @ 2hrs/ w/30lbs
WEEK 2				
Circuit Training (CT)	RT	ABGR: 4-mi UBS	RT	Unit Run
WEEK 3				
ABGR: 4-mi	RT	ABGR: 4-mi	Anaerobic Training (ANT): 6x 100m sprints w/ 45-sec rest	Hike: 11km/6.8-mi <2.75hrs w/40lbs
WEEK 4				
RT	ABGR: 5-mi	CT	RT	Unit Run
WEEK 5				
ABGR: 5-mi UBS	RT	ABGR: 5-mi	ANT: 8x 100m sprints w/ 45-sec rest	Hike: 13km/8-mi <3.25hrs w/50lbs
WEEK 6				
PFT 4-mi	RT	ABGR: 6-mi UBS	RT	Unit Run

d. <u>Conditioning Program "A"</u>. Weekly aerobic program to address problems in the PFT and other distance running.

<u>Monday</u>: Distance/Time Run 30-minute run (slower pace than PFT run pace)

<u>Tuesday</u>: Cross training, Conduct cardiovascular training other than running; e.g., circuit training, swimming, or biking

<u>Wednesday</u>: Interval Training

Interval runs on track or known distance course - 4x 400 sprints;
- Increase # of sprints with conditioning
- Increase sprint distance with conditioning

<u>Thursday</u>: Cross Training

<u>Friday</u>: Distance Run - increase duration/speed from Monday's run

e. <u>Conditioning Program "B" (6 months)</u>. Program is designed to address PFT/distance running weaknesses, using a long-term (6 months) conditioning period. Run training should be conducted three times per week on non-consecutive days (e.g., Monday/Wednesday/Friday). One of the runs will be an interval run and the remaining two should be distance runs. The following schedule depicts the interval runs for every week for the 6-month period.

INTERVAL RUN TRAINING

WEEK	RUN DIST (YDS:MI)	MAX TIME (MIN:SEC)	WALK DIST: TIME (YDS:SEC)	REPS-RUN/WALK	TOTAL TIME (MIN:SEC)	TOTAL MILES
1,2	220: 1/8	1:00	110:60	11	22:00	2
3,4	330:	1:30	110:60	9	22:30	2.25
5,6	440: 1/4	2:20	110:75	8	26:40	2.5
7,8	660:	4:00	110:90	6	31:00	2.63
9,10	880: 1/2	5:30	110:120	3	32:45	2.63
	440: 1/4	2:10	110:75	3		
11,12	880: 1/2	5:20	110:120	4	35:50	2.81
	440: 1/4	2:00	110:75	2		
13,14	880: 1/2	5:20	110:105	4	38:20	3.06
	660:	3:45	110:75	2		
15,16	880: 1/2	5:00	110:105	5	36:00	3.19
	660:	3:30	110:75	1		
17,18	880: 1/2	4:45	110:90	6	37:30	3.38
19,20	1 M 4 laps	9:00	110:120	1	35:00	3.31
	880: 1/2	4:30	110:90	4		
21-24	1 M 4 laps	8:30	110:105	1	33:15	3.31
	880: 1/2	4:15	110:90	4		

The following chart shows the distance runs (per training session) for every week for the 6-month period. The distance runs should be conducted at 75% of maximum effort.

DISTANCE RUN TABLE

WEEKS	RUN DISTANCE (MILES)
1-4	2.5
5-8	2.75
9-12	3.0
13-16	3.25
17-20	3.5
21-24	4.0

Each PT session should include a substantial warmup, stretching, and cool-down from the Daily 16.

f. Conditioning Program "C" (Single Activity). The "single activity" conditioning schedule provides a 60-minute training session in which the single activity system is used with different events each day. In this program, 5 hours of training are scheduled for the week, and supplemented with integrated unit training each week. A bimonthly conditioning march of 10 miles in 3 hours is suggested to round out this program. Such a program could be used as a transition period (1-4 weeks) between other remedial/conditioning programs.

DAY	TIME	ACTIVITY
Monday	60-min	Running Activity
Tuesday	60-min	Combatives
Wednesday	60-min	Conditioning & Log Drills
Thursday	60-min	Conditioning & Relays
Friday	60-min	Running & Conditioning
Saturday (optional)		Running Activity

g. Conditioning Program "D" (Single Activity). This conditioning program schedule illustrates the use of another "single activity" system for 5.25 hours of training, to include sustaining-type activities for a unit that has passed through both the toughening and slow improvement stages of conditioning. The use of competitive activities is featured to hold interest and provide self-motivation. A bimonthly 10-mile/3-hour march under load will be necessary to round out this program.

DAY	TIME	ACTIVITY
Monday	45-min	Running - Terrain Running
Tuesday	75-min	Team Sports or Ability Group Run/Interval Sets
Wednesday	45-min	Running - Fartlek or Distance Run
Thursday	20-min 40-min	Log Drills Dual Combatives
Friday	90-min	Speed March
Saturday (optional)		Team Sports/Team Contests

h. Conditioning Program "E" (Multi-Activities). This program provides a more comprehensive training regimen of various training periods and activities, and the use of three systems during the training week. In this schedule, 3.5 hours are included for the week.

DAY	TIME	ACTIVITY
Monday	20-min 20-min 20-min	Rifle Drill Combatives Running
Tuesday	30-min	Daily 16 Grass Drills
Wednesday	30-min	Rifle Drill
Thursday	30-min	Circuit Interval Training
Friday	20-min 20-min 20-min	Strength Circuit Conditioning Drills Cross-Country Run
Saturday (optional)	60-min	Warmup Run/Grass Drills

On Monday, a 60-min period is available for PT with three, 20-min segments. Groups will progress through each scheduled activity in turn. On Tuesday, Wednesday, and Thursday, only 30-min periods are available, conducted by the entire unit. As with the Monday schedule, on Friday, the unit should conduct the 20-min segments as subgroups, again progressing through each scheduled activity in turn. Bimonthly conditioning marches will be necessary to augment the program.

APPENDIX D

BODY COMPOSITION PROGRAM (BCP) ASSIGNMENT PROCESS

LETTERHEAD

6100
Office Code
Date

From: Commanding Officer
To: Medical Officer

Subj: BODY COMPOSITION PROGRAM (BCP) EVALUATION ON RANK SNM
XXX XX XXX/XXXX

Ref: (a) MCO P6100.12

1. It has been determined that (RANK SNM)____________________________ is not within the Corps' body composition standards.

2. During the semi-annual body composition evaluation conducted on (YYMMDD) ____________________, SNM was found to weigh ________ pounds and has a ________ percent body fat. Per the reference, the Marine's maximum weight is ________ pounds based on a height of ________ inches.

3. SNM received a Physical Fitness Test (PFT) score on (YYMMDD) ________________, which is a ($1^{st}/2^{nd}/3^{rd}$) ________ Class score.

4. SNM was screened and does not meet the (PFT performance/percent body fat) __________________________ criteria and is therefore not eligible for the Physical Performance Evaluation.

5. SNM has been advised that the loss of ______ pounds or _____ percent body fat is required and this reduced weight/body fat percentage must be maintained within the 6-month BCP assignment period.

6. Request a medical evaluation, per the reference, be conducted to determine SNM's medical status for BCP and Remedial Physical Conditioning Program (RPCP) participation.

____
CO's Signature

Copy to:

ENCLOSURE (1)

APPENDIX D

BODY COMPOSITION PROGRAM (BCP) ASSIGNMENT PROCESS
LETTERHEAD

6100
Office Code
Date

FIRST ENDORSEMENT on (identify letter)

From: Medical Officer
To: Commanding Officer

Subj: BODY COMPOSITION PROGRAM (BCP) EVALUATION ON RANK SNM
XXX XX XXXX/XXXX

Ref: (a) MCO P6100.12

1. Returned.

2. Per the reference, it has been determined that (RANK SNM) ____________________ is not within body composition standards for the following (initial one of the following):

a. (_____) Due to an underlying cause or associated disease. Attached are the recommended follow-on medical tests/evaluations.

b. (_____) Due to an underlying cause or associated disease. Attached is the prescribed treatment for the condition. The Residential Obesity Program is/is not recommended for this individual.

c. (_____) Not to an underlying cause or associated disease. SNM is fit for participation in a BCP and Remedial Physical Conditioning Program (RPCP). Attached are the prescribed weight and body fat reduction goals and dietary guidelines.

d. (_____) Not to an underlying cause or associated disease. SNM is fit for participation in a BCP and alternative RPCP. Attached are the physical restrictions/limitations imposed, prognosis for return to a full duty status, weight and body fat reduction goals, and dietary guidelines.

____
MO's (ACHCP's) Signature

____
Grade

____
Branch of Service

____
Medical Title

____
Organization

APPENDIX D

BODY COMPOSITION PROGRAM (BCP) ASSIGNMENT PROCESS
LETTERHEAD

6100
Office Code
Date

SECOND ENDORSEMENT on (identify letter)

From: Commanding Officer
To: Individual Marine

Subj: BODY COMPOSITION PROGRAM (BCP) EVALUATION ON RANK SNM
XXX XX XXXX/XXXX

Ref: (a) MCO P6100.12
(b) MCO P1080.40C
(c) MCO P1900.16F

1. Readdressed and forwarded.

2. It has been determined that you are not in adherence with Marine Corps body composition standards and are not eligible for the Physical Performance Evaluation. You are being assigned to a 6-month BCP as a ($1^{st}/2^{nd}$) ________ assignment. You are directed to meet the following weight reduction goal of ______ pounds or a body fat reduction goal of ______ percent per month. Your goal is to lose ______ pounds and ______ percent body fat.

3. Your mandatory participation in the BCP and following Program begins on (YYMMDD) _______________ and ends on (YYMMDD) _______________.

a. (_____) The Unit's Remedial Physical Conditioning Program (RPCP) for 6 months per reference (a).

b. (_____) The alternate RPCP as a result of physical restrictions/ limitations imposed by the Medical Officer per reference (a).

4. You are advised that assistance in meeting and maintaining your weight and body fat reduction goals is available through

____

____

____.

5. Official assignment to the 6-month BCP will be reported on the Unit Diary per reference (b).

6. You are advised that failure to comply with reference (a) may result in administrative separation per paragraph 6206 or 6215 of reference (c) respectively.

____
CO's Signature

____
Unit Diary Number

____
Date

APPENDIX D

BODY COMPOSITION PROGRAM (BCP) ASSIGNMENT PROCESS
LETTERHEAD

6100
Office Code
Date

THIRD ENDORSEMENT on (identify letter)

From: Individual Marine
To: Commanding Officer

Subj: BODY COMPOSITION PROGRAM (BCP) EVALUATION ON RANK SNM
XXX XX XXXX/XXXX

1. Returned with acknowledgments. (SNM must read and initial all paragraphs below.)

2. (_____) I understand I am not in adherence with Marine Corps body composition standards, and it is not due to an underlying cause or associated disease, based on the medical evaluation completed by a Medical Officer (MO).

3. (_____) I understand I am not eligible for the Physical Performance Evaluation and I am being assigned to the BCP, as a ($1^{st}/2^{nd}$) ________ assignment.

4. (_____) I understand I am being directed to meet the following weight reduction goal of ______ pounds or a body fat reduction goal of ______ percent per month, and my weight goal is ______ pounds and body fat goal is ______ percent.

5. I understand my responsibility to (initial all):

a. (_____) Participate in the Unit's BCP for a total of 6 months.

b. (_____) Participate in the Unit Remedial Physical Conditioning Program (RPCP) or alternate RPCP, as prescribed by the evaluating MO for the duration of my 6-month BCP assignment.

c. (_____) Have my weight and body fat measurements taken and recorded once every 2 weeks by the Command Physical Training Representative.

d. (_____) Follow the dietary regimen prescribed by the evaluating MO.

e. (_____) Complete the MCI Course 3316, "Basic Nutrition." (First assignment to BCP only).

____
Individual Marine's Signature

BODY COMPOSITION PROGRAM (BCP) ASSIGNMENT PROCESS
LETTERHEAD

6100
Office Code
Date

FOURTH ENDORSEMENT on (identify letter)

From: Commanding Officer
To: Medical Officer

Subj: BODY COMPOSITION PROGRAM (BCP) EVALUATION ON RANK SNM
XXX XX XXXX/XXXX

1. Readdressed and forwarded for action.

2. SNM has been participating in the BCP and Remedial Physical Conditioning Program for 4 months. A medical reevaluation is requested on SNM, based on progress made towards weight and body fat reduction goals. If weight and body fat reduction goals are not being met or maintained, verify the diagnosis of no underlying cause or associated disease.

3. SNM has lost ______ pound(s) and ______ percent body fat, since being placed to the BCP on (YYMMDD) ______________.

____
CO's Signature

APPENDIX D

BODY COMPOSITION PROGRAM (BCP) ASSIGNMENT PROCESS
LETTERHEAD

6100
Office Code
Date

FIFTH ENDORSEMENT on (identify letter)

From: Medical Officer
To: Commanding Officer

Subj: BODY COMPOSITION PROGRAM (BCP) EVALUATION ON RANK SNM
XXX XX XXXX/XXXX

Ref: (a) MCO P6100.12

1. Returned.

2. SNM has been reevaluated, resulting in the following diagnosis/ recommendation: (Initial appropriate paragraph)

a. (_____) SNM's current body composition status is due to an underlying cause or associated disease. Attached are the diagnosis and recommended follow-on actions to be taken per paragraph 3200 of the reference.

b. (_____) SNM's present body composition status is not due to an underlying cause or associated disease. Participation in the BCP and Remedial Physical Conditioning Program should continue until Program assignment expires.

3. Based on the above diagnoses, medical comments and recommendations are attached.

____
MO's (ACHCP's) Signature

____
Grade

____
Branch of Service

____
Medical Title

____
Organization

APPENDIX D

BODY COMPOSITION PROGRAM (BCP) ASSIGNMENT PROCESS
LETTERHEAD

6100
Office Code
Date

SIXTH ENDORSEMENT on (identify letter)

From: Commanding Officer
To: Individual Marine

Subj: BODY COMPOSITION PROGRAM (BCP) EVALUATION ON RANK SNM
XXX XX XXXX/XXXX

Ref: (a) MCO P1080.40C
(b) MCO P1900.16F

1. Readdressed and forwarded.

2. Your ($1^{st}/2^{nd}$) ________ 6-month assignment to the BCP expired on (YYMMDD) ________________.

3. Based on your progress while assigned to the BCP, as documented in your BCP file, it has been determined that: (Initial one)

a. (_____) You have attained and maintained the Marine Corps' body composition standards, and will be officially removed from the BCP (and Remedial Physical Conditioning Program (RPCP)) on the Unit Diary per reference (a).

(1) (_____) FIRST ASSIGNMENT PARTICIPANTS. If at any time during your Marine Corps career you fail to meet the body composition standards a second time, you may be granted a second BCP (and RPCP) assignment of 6 months with no extension.

(2) (_____) SECOND ASSIGNMENT PARTICIPANTS. If at any time during your Marine Corps career you fail to meet the body composition standards a third time, you will automatically be processed for administrative separation per reference (b).

b. (_____) FIRST ASSIGNMENT PARTICIPANTS ONLY. You have not met the Marine Corps' body composition standards, but have made satisfactory progress during your first 6-month BCP assignment. You have been granted an extension for ______ additional months, and will be required to return to body composition standards upon Program extension expiration on (YYMMDD) _______________. This extension has been entered on the Unit Diary per reference (a).

(1) You will continue your participation in the BCP and RPCP throughout the extension period.

(2) Thirty-days prior to Program expiration, you will be medically reevaluated.

(3) Upon Program expiration, if body composition standards are met, you will be officially removed from the BCP (and RPCP). If body composition standards are not met, you will be processed for administrative separation per paragraph 6215 in reference (b).

BODY COMPOSITION PROGRAM (BCP) ASSIGNMENT PROCESS

SIXTH ENDORSEMENT on (identify letter)

Subj: BODY COMPOSITION PROGRAM (BCP) EVALUATION ON RANK SNM
XXX XX XXXX/XXXX

c. (_____) You have failed to meet the Marine Corps' body composition standards and will receive a 6105 counseling entry on your SRB/OQR's Page 11, and be processed for administrative separation per paragraph 6206 or 6215 in reference (b) respectively.

____
CO's Signature

____
Individual Marine's Signature

____
Unit Diary Number

____
Date

LETTERHEAD

6100
Office Code
Date

SEVENTH ENDORSEMENT on (identify letter)

From: Commanding Officer
To: Individual Marine

Subj: BODY COMPOSITION PROGRAM (BCP) EVALUATION ON RANK SNM
XXX XX XXXX/XXXX

Ref: (a) MCO P1080.40C
(b) MCO P1900.16F

1. Readdressed and forwarded.

2. Your extension to your first 6-month BCP assignment expired on (YYMMDD) ________________.

3. Based on your progress while assigned to the BCP, as documented in your BCP file, it has been determined that: (Initial one)

a. (_____) You have attained and maintained the Marine Corps' body composition standards, and will be officially removed from the BCP (and Remedial Physical Conditioning Program (RPCP)) on the Unit Diary per reference (a). If at any time during your Marine Corps career you fail to meet the body composition standards a second time, you may be granted a second BCP (and RPCP) assignment of 6-months with no extension.

b. (_____) You have failed to meet the Marine Corps' body composition standards and will receive a 6105 counseling entry on your SRB/OQR's Page 11, and be processed for administrative separation per paragraph 6206 or 6215 in reference (b) respectively.

____
CO's Signature

____
Individual Marine's Signature

____
Unit Diary Number

____
Date

APPENDIX D

PHYSICAL PERFORMANCE EVALUATION FLOWCHART

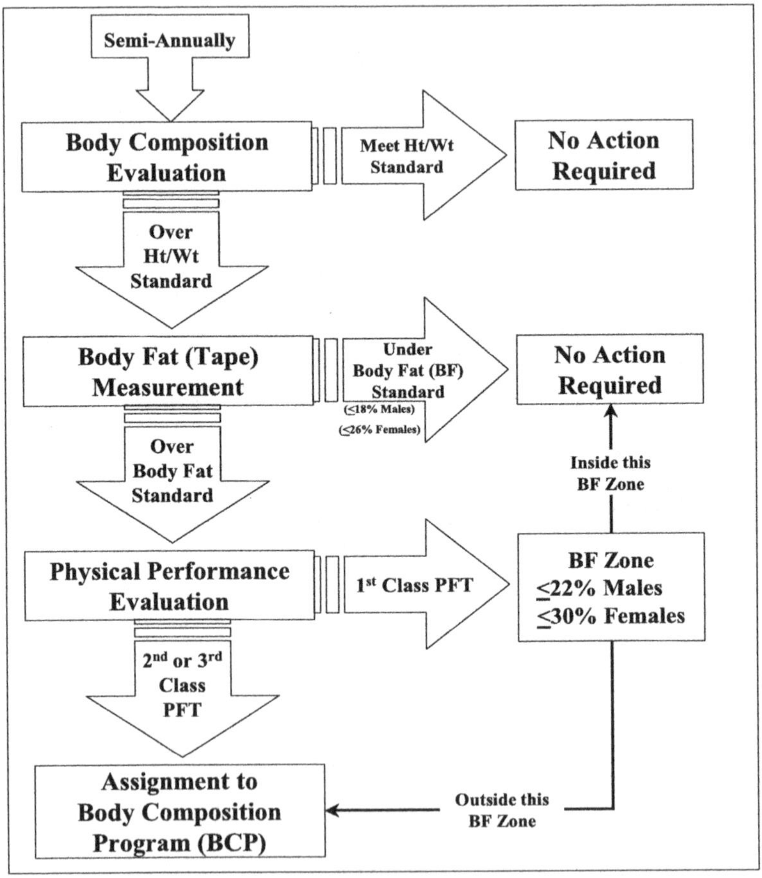

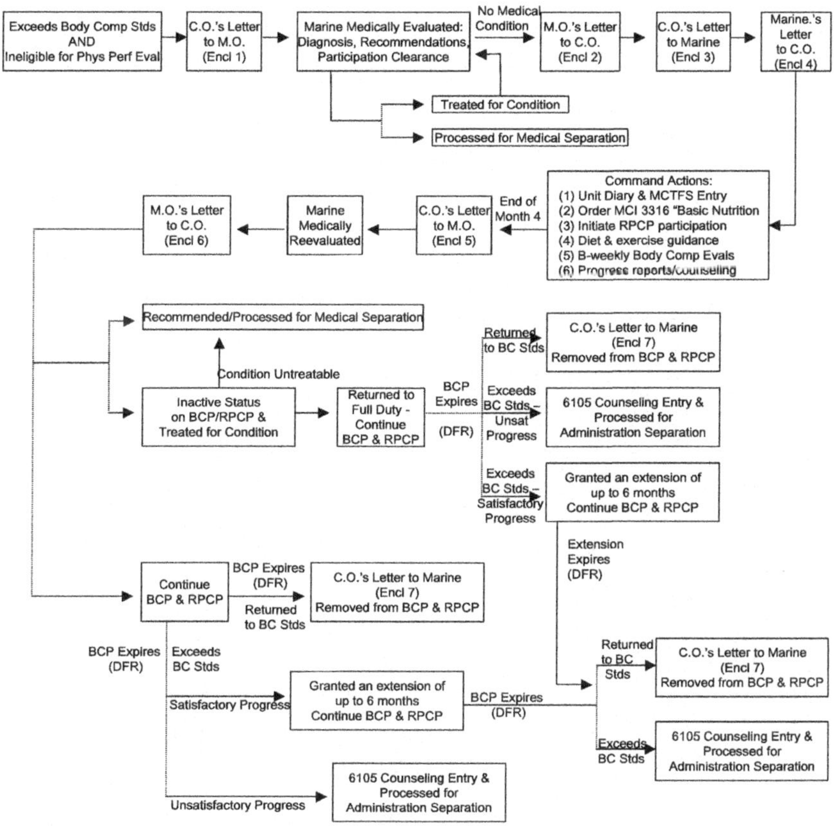

Appendix D: Body Composition Program Assignment Process

NOTES:

(1) Marines assigned to a Body Composition Program (BCP) will also be assigned to the Remedial Physical Conditioning Program (RPCP) for the duration of the 6-month BCP Assignment/Extension.

(2) A Marine on light/limited duty will participate in an alternative RPCP.

(3) 2nd-Assignment Marines, who successfully completed their 1st Assignment and fail to meet BC standards later in their career, will be given 6 months (with no estension) for a 2nd Assignment.

(4) No Third Chances: Marines failing to meet BC standards a third time in their career will be immediately processed for administrative separation, and will informally participate in the BCP & RPCP until discharged. No 3rd Assignment will be given under any circumstances.

(5) Marines will receive a 6105 counseling entry on their SRB/OQR's Page 11 per paragraph 6206 or 6215 of MCO P1900.16F respectively.

APPENDIX E

AGE 46 AND OVER RISK FACTOR SCREENING CHECKLIST

AGE 46 AND OVER RISK FACTOR SCREENING

NAME (LAST, FIRST, M.I.):		SSN:
COMMAND:	SECTION:	PHONE:

SECTION A: (COMPLETED BY THE MARINE)	Circle
(1) ARE YOU NOT IN ACCORDANCE WITH MEDICAL PHYSICAL EXAMINATION REQUIREMENTS FOR YOUR AGE?	YES NO
(2) DATE OF LAST PERIODIC PHYSICAL EXAMINATION:	YES NO
SECTION B: (COMPLETED BY THE MARINE)	
(1) DO YOU HAVE A HISTORY OF HEART DISEASE OR HIGH BLOOD PRESSURE WHICH REQUIRES YOU TO RESTRICT PHYSICAL ACTIVITY OR SEEK MEDICAL TREATMENT?	YES NO
(2) SINCE YOUR LAST PERIODIC PHYSICAL EXAMINATION, SEMI-ANNUAL RISK FACTOR SCREENING, OR PFT, HAS THERE BEEN:	
A. A CHANGE IN YOUR HISTORY OF HEART DISEASE OR HIGH BLOOD PRESSURE WHICH REQUIRED YOU TO RESTRICT PHYSICAL ACTIVITY OR SEEK MEDICAL TREATMENT?	YES NO
B. A CHANGE IN YOUR FAMILY HISTORY SUCH AS IMMEDIATE FAMILY MEMBERS EXPERIENCING A HEART ATTACK OR DIED OF HEART DISEASE BEFORE THEY WERE 45 YEARS OLD?	YES NO
C. A CHANGE IN YOUR INCIDENCE OF DISCOMFORT TO YOUR CHEST, ARMS, OR NECK WHILE EXERCISING?	YES NO
D. AN INCIDENT WHERE YOU WERE PRONE TO HEAT EXHAUSTION/HEAT STROKE, FEELING FAINT, OR FEELING THAT YOU WERE ABOUT TO LOSE CONSCIOUSNESS?	YES NO
E. A CHANGE IN ANY MEDICAL CONDITION (E.G., DIABETES, ASTHMA, BONE OR JOINT DISEASE) WHICH YOU THINK MIGHT LIMIT YOUR PARTICIPATION IN THE PFT?	YES NO
F. A SIGNIFICANT WEIGHT CHANGE?	YES NO
G. A CHANGE IN YOUR SMOKING HABITS SUCH THAT YOU NOW SMOKE ONE OR MORE PACKS OF CIGARETTES DAILY?	YES NO

MARINE'S SIGNATURE : DATE:

SECTION C: (COMPLETED BY THE MEDICAL OFFICER (M.O.) IF REQUIRED)

DATE OF MEDICAL EVALUATION: ______________________

(1) THE MARINE'S PERIODIC PHYSICAL EXAMINATION IS CURRENT (Y/N)? __________
REMARKS:____

(2) THE MARINE ANSWERED 'YES' TO QUESTIONS IN SECTION (B)(Y/N)? ________
REMARKS:____

SECTION D: (COMPLETED BY THE MEDICAL OFFICER (M.O.) IF REQUIRED)

(1) MEDICAL RECOMMENDATION: REFERRED FOR FURTHER MEDICAL CONSULTATION (Y/N)? _____ (E.G., PSYCHOLOGY, NUTRITION, INTERNAL MEDICINE, PHYSICAL THERAPY)
REMARKS:____

(2) CLEARED FOR PARTICIPATION IN PFT (Y/N)? _____
REMARKS:____

SIGNATURE OF M.O. DATE:

PRINTED NAME/RANK/COMMAND/UNIT/PHONE:

APPENDIX F

PHYSICAL FITNESS TEST AND BODY COMPOSITION WORKSHEET

UNIT:

DATE:

MONITOR:

Remarks

* %BF required ONLY if Marine exceeds Ht/Wt standards.

** Semiannual Risk Factor Screening required ONLY if Marine is age 46 and over.

NAME/SSN	RANK	AGE	GENDER M/F	HT (in.)	WT (lbs.)	% BF*	Risk Factor Screening**	Pullups Fix Arm Hang	SCORE	Crunches	SCORE	3-Mile Run	SCORE	PFT TOTAL	Partial PFT TOTAL	CLASS	Phys Perf Eval (Y/N)

APPENDIX G

PHYSICAL FITNESS TEST SCORING TABLE

Sec I: FEMALES

Points	Flexed-Arm Hang	Crunches	3-Mile Run
100	70 sec	100	21:00
99		99	21:10
98	69 sec	98	21:20
97		97	21:30
96	68 sec	96	21:40
95		95	21:50
94	67 sec	94	22:00
93		93	22:10
92	66 sec	92	22:20
91		91	22:30
90	65 sec	90	22:40
89		89	22:50
88	64 sec	88	23:00
87		87	23:10
86	63 sec	86	23:20
85		85	23:30
84	62 sec	84	23:40
83		83	23:50
82	61 sec	82	24:00
81		81	24:10
80	60 sec	80	24:20
79		79	24:30
78	59 sec	78	24:40
77		77	24:50
76	58 sec	76	25:00
75		75	25:10
74	57 sec	74	25:20
73		73	25:30
72	56 sec	72	25:40
71		71	25:50
70	55 sec	70	26:00
69		69	26:10
68	54 sec	68	26:20
67		67	26:30
66	53 sec	66	26:40
65		65	26:50
64	52 sec	64	27:00
63		63	27:10
62	51 sec	62	27:20
61		61	27:30
60	50 sec	60	27:40
59		59	27:50
58	49 sec	58	28:00
57		57	28:10
56	48 sec	56	28:20
55		55	28:30
54	47 sec	54	28:40
53		53	28:50
52	46 sec	52	29:00
51		51	29:10

Points	Flexed-Arm Hang	Crunches	3-Mile Run
50	45 sec	50	29:20
49		49	29:30
48	44 sec	48	29:40
47		47	29:50
46	43 sec	46	30:00
45		45	30:10
44	42 sec	44	30:20
43		43	30:30
42	41 sec	42	30:40
41		41	30:50
40	40 sec	40	31:00
39	39 sec	x	31:10
38	38 sec	x	31:20
37	37 sec	x	31:30
36	36 sec	x	31:40
35	35 sec	x	31:50
34	34 sec	x	32:00
33	33 sec	x	32:10
32	32 sec	x	32:20
31	31 sec	x	32:30
30	30 sec	x	32:40
29	29 sec	x	32:50
28	28 sec	x	33:00
27	27 sec	x	33:10
26	26 sec	x	33:20
25	25 sec	x	33:30
24	24 sec	x	33:40
23	23 sec	x	33:50
22	22 sec	x	34:00
21	21 sec	x	34:10
20	20 sec	x	34:20
19	19 sec	x	34:30
18	18 sec	x	34:40
17	17 sec	x	34:50
16	16 sec	x	35:00
15	15 sec	x	35:10
14	x	x	35:20
13	x	x	35:30
12	x	x	35:40
11	x	x	35:50
10	x	x	36:00
9	x	x	x
8	x	x	x
7	x	x	x
6	x	x	x
5	x	x	x
4	x	x	x
3	x	x	x
2	x	x	x
1	x	x	x

***Round up all values (e.g., 21:01 to 21:09 equals 99 points)**

APPENDIX G

PHYSICAL FITNESS TEST SCORING TABLE

Sec II:	MALES						
Points	Pull-ups	Crunches	3-Mile Run	**Points**	Pull-ups	Crunches	3-Mile Run
100	20	100	18:00	**50**	10	50	26:20
99		99	18:10	**49**		49	26:30
98		98	18:20	**48**		48	26:40
97		97	18:30	**47**		47	26:50
96		96	18:40	**46**		46	27:00
95	19	95	18:50	**45**	9	45	27:10
94		94	19:00	**44**		44	27:20
93		93	19:10	**43**		43	27:30
92		92	19:20	**42**		42	27:40
91		91	19:30	**41**		41	27:50
90	18	90	19:40	**40**	8	40	28:00
89		89	19:50	**39**		x	28:10
88		88	20:00	**38**		x	28:20
87		87	20:10	**37**		x	28:30
86		86	20:20	**36**		x	28:40
85	17	85	20:30	**35**	7	x	28:50
84		84	20:40	**34**		x	29:00
83		83	20:50	**33**		x	29:10
82		82	21:00	**32**		x	29:20
81		81	21:10	**31**		x	29:30
80	16	80	21:20	**30**	6	x	29:40
79		79	21:30	**29**		x	29:50
78		78	21:40	**28**		x	30:00
77		77	21:50	**27**		x	30:10
76		76	22:00	**26**		x	30:20
75	15	75	22:10	**25**	5	x	30:30
74		74	22:20	**24**		x	30:40
73		73	22:30	**23**		x	30:50
72		72	22:40	**22**		x	31:00
71		71	22:50	**21**		x	31:10
70	14	70	23:00	**20**	4	x	31:20
69		69	23:10	**19**		x	31:30
68		68	23:20	**18**		x	31:40
67		67	23:30	**17**		x	31:50
66		66	23:40	**16**		x	32:00
65	13	65	23:50	**15**	3	x	32:10
64		64	24:00	**14**	x	x	32:20
63		63	24:10	**13**	x	x	32:30
62		62	24:20	**12**	x	x	32:40
61		61	24:30	**11**	x	x	32:50
60	12	60	24:40	**10**	x	x	33:00
59		59	24:50	**9**	x	x	x
58		58	25:00	**8**	x	x	x
57		57	25:10	**7**	x	x	x
56		56	25:20	**6**	x	x	x
55	11	55	25:30	**5**	x	x	x
54		54	25:40	**4**	x	x	x
53		53	25:50	**3**	x	x	x
52		52	26:00	**2**	x	X	x
51		51	26:10	**1**	x	x	x

*** Round up all values (e.g., 18:01 to 18:09 equals 99 points)**

APPENDIX H

ALTITUDE COMPENSATION FOR 3.0 MILE PHYSICAL FITNESS RUN CHART

Points	3.0-Mile (Male)	3.0-Mile (Female)	Points	3.0-Mile (Male)	3.0-Mile (Female)	Points	3.0-Mile (Male)	3.0-Mile (Female)
100	19:30	22:30	**65**	25:20	:20	**30**	31:10	34:10
99	199:40	22:40	**64**	25:30	28:30	**29**	31:20	34:20
98	19:50	22:50	**63**	25:40	28:40	**28**	31:30	34:30
97	20:00	23:00	**62**	25:50	28:50	**27**	31:40	34:40
96	20:10	23:10	**61**	26:00	29:00	**26**	31:50	34:50
95	20:20	23:20	**60**	26:10	29:10	**25**	32:00	35:00
94	20:30	23:30	**59**	26:20	29:20	**24**	32:10	35:10
93	20:40	23:40	**58**	26:30	29:30	**23**	32:20	35:20
92	20:50	23:50	**57**	26:40	29:40	**22**	32:30	35:30
91	21:00	24:00	**56**	26:50	29:50	**21**	32:40	35:40
90	21:10	24:10	**55**	27:00	30:00	**20**	32:50	35:50
89	21:20	24:20	**54**	27:10	30:10	**19**	33:00	36:00
88	21:30	24:30	**53**	27:20	30:20	**18**	33:10	36:10
87	21:40	24:40	**52**	27:30	30:30	**17**	33:20	36:20
86	21:50	24:50	**51**	27:40	30:40	**16**	33:30	36:30
85	22:00	25:00	**50**	27:50	30:50	**15**	33:40	36:40
84	22:10	25:10	**49**	28:00	31:00	**14**	33:50	36:50
83	22:20	25:20	**48**	28:10	31:10	**13**	34:00	37:00
82	22:30	25:30	**47**	28:20	31:20	**12**	34:10	37:10
81	22:40	25:40	**46**	28:30	31:30	**11**	34:20	37:20
80	22:50	25:50	**45**	28:40	31:40	**10**	34:30	37:30
79	23:00	26:00	**44**	28:50	31:50	**9**	34:40	37:40
78	23:10	26:10	**43**	29:00	32:00	**8**	34:50	37:50
77	23:20	26:20	**42**	29:10	32:10	**7**	35:00	38:00
76	23:30	26:30	**41**	29:20	32:20	**6**	35:10	38:10
75	23:40	26:40	**40**	29:30	32:30	**5**	35:20	38:20
74	23:50	26:50	**39**	29:40	32:40	**4**	35:30	38:30
73	24:00	27:00	**38**	29:50	32:50	**3**	35:40	38:40
72	24:10	27:10	**37**	30:00	33:00	**2**	35:50	38:50
71	24:20	27:20	**36**	30:10	33:10	**1**	36:00	39:00
70	24:30	27:30	**35**	30:20	33:20			
69	24:40	27:40	**34**	30:30	33:30			
68	24:50	27:50	**33**	30:40	33:40			
67	25:00	28:00	**32**	30:50	33:50			
66	25:10	28:10	**31**	31:00	34:00			

APPENDIX I

SECTION 1: HEIGHT/WEIGHT STANDARDS

MALES

HEIGHT (Inches)	Maximum Standard (Pounds)	Minimum Standard (Pounds)
58"	132	91
59"	136	94
60"	141	97
61"	146	100
62"	150	104
63"	155	107
64"	160	110
65"	165	114
66"	170	117
67"	176	121
68"	181	125
69"	186	128
70"	192	132
71"	197	136
72"	203	140
73"	208	144
74"	214	148
75"	220	152
76"	226	156
77"	232	160
78"	238	164
79"	244	168
80"	250	173

Max BF%:	18%

FEMALES

HEIGHT (Inches)	Maximum Standard (Pounds)	Minimum Standard (Pounds)
58"	120	91
59"	124	94
60"	128	97
61"	132	100
62"	137	104
63"	141	107
64"	146	110
65"	150	114
66"	155	117
67"	160	121
68"	164	125
69"	169	128
70"	174	132
71"	179	136
72"	184	140
73"	189	144
74"	195	148
75"	200	152
76"	205	156
77"	211	160
78"	216	164
79"	222	168
80"	228	173

Max BF%:	26%

SECTION 2: PERCENT BODY FAT ESTIMATION

PERCENT BODY FAT ESTIMATION FOR MEN

Circumference Value*	Height (in)									
	60.0	60.5	61.0	61.5	62.0	62.5	63.0	63.5	64.0	64.5
13.5	9	9								
14.0	11	11	10	10	10	10	9	9		
14.5	12	12	12	11	11	11	11	10	10	10
15.0	13	13	13	13	12	12	12	12	11	11
15.5	15	14	14	14	14	13	13	13	13	12
16.0	16	16	15	15	15	15	14	14	14	14
16.5	17	17	16	16	16	16	15	15	15	15
17.0	18	18	18	17	17	17	17	16	16	16
17.5	19	19	19	18	18	18	18	17	17	17
18.0	20	20	20	19	19	19	19	18	18	18
18.5	21	21	21	20	20	20	20	19	19	19
19.0	22	22	22	21	21	21	21	20	20	20
19.5	23	23	23	22	22	22	22	21	21	21
20.0	24	24	24	23	23	23	23	22	22	22
20.5	25	25	25	24	24	24	24	23	23	23
21.0	26	26	25	25	25	25	24	24	24	24
21.5	27	27	26	26	26	26	25	25	25	25
22.0	28	27	27	27	27	26	26	26	26	25
22.5	29	28	28	28	28	27	27	27	27	26
23.0	29	29	29	29	28	28	28	28	27	27
23.5	30	30	30	29	29	29	29	28	28	28
24.0	31	31	30	30	30	30	29	29	29	29
24.5	32	31	31	31	31	30	30	30	30	29
25.0	32	32	32	32	31	31	31	31	30	30
25.5	33	33	33	32	32	32	32	31	31	31
26.0	34	34	33	33	33	33	32	32	32	32
26.5	35	34	34	34	34	33	33	33	33	32
27.0	35	35	35	35	34	34	34	34	33	33
27.5	36	36	36	35	35	35	35	34	34	34
28.0	37	36	36	36	36	35	35	35	35	34
28.5			37	37	36	36	36	36	35	35
29.0					37	37	37	36	36	36
29.5								37	37	36
30.0										
30.5										
31.0										
31.5										
32.0										
32.5										
33.0										
33.5										
34.0										
34.5										
35.0										

* Circumference Value = abdomen circumference - neck circumference (in inches)

SECTION 2: PERCENT BODY FAT ESTIMATION

PERCENT BODY FAT ESTIMATION FOR MEN

Circumference Value*	Height (in)									
	65.0	65.5	66.0	66.5	67.0	67.5	68.0	68.5	69.0	69.5
13.5										
14.0										
14.5	10	9	9							
15.0	11	11	10	10	10	10	10	9	9	
15.5	12	12	12	11	11	11	11	11	10	10
16.0	13	13	13	13	12	12	12	12	12	11
16.5	14	14	14	14	14	13	13	13	13	12
17.0	16	15	15	15	15	14	14	14	14	14
17.5	17	16	16	16	16	16	15	15	15	15
18.0	18	18	17	17	17	17	16	16	16	16
18.5	19	19	18	18	18	18	17	17	17	17
19.0	20	20	19	19	19	19	18	18	18	18
19.5	21	21	20	20	20	20	19	19	19	19
20.0	22	21	21	21	21	21	20	20	20	20
20.5	23	22	22	22	22	21	21	21	21	21
21.0	24	23	23	23	23	22	22	22	22	21
21.5	24	24	24	24	23	23	23	23	23	22
22.0	25	25	25	25	24	24	24	24	23	23
22.5	26	26	26	25	25	25	25	24	24	24
23.0	27	27	26	26	26	26	26	25	25	25
23.5	28	27	27	27	27	27	26	26	26	26
24.0	28	28	28	28	28	27	27	27	27	26
24.5	29	29	29	29	28	28	28	28	27	27
25.0	30	30	30	29	29	29	29	28	28	28
25.5	31	31	30	30	30	30	29	29	29	29
26.0	31	31	31	31	31	30	30	30	30	29
26.5	32	32	32	32	31	31	31	31	30	30
27.0	33	33	32	32	32	32	32	31	31	31
27.5	34	33	33	33	33	32	32	32	32	32
28.0	34	34	34	34	33	33	33	33	32	32
28.5	35	35	34	34	34	34	34	33	33	33
29.0	36	35	35	35	35	34	34	34	34	34
29.5	36	36	36	36	35	35	35	35	34	34
30.0	37	37	36	36	36	36	35	35	35	35
30.5			37	37	37	36	36	36	36	35
31.0						37	37	36	36	36
31.5									37	37
32.0										
32.5										
33.0										
33.5										
34.0										
34.5										
35.0										

* Circumference Value = abdomen circumference - neck circumference (in inches)

APPENDIX I

SECTION 2: PERCENT BODY FAT ESTIMATION

PERCENT BODY FAT ESTIMATION FOR MEN

Circumference Value*	Height (in)									
	70.0	70.5	71.0	71.5	72.0	72.5	73.0	73.5	74.0	74.5
13.5										
14.0										
14.5										
15.0										
15.5	10	10	9	9	9					
16.0	11	11	11	10	10	10	10	10	9	9
16.5	12	12	12	12	11	11	11	11	11	10
17.0	13	13	13	13	13	12	12	12	12	11
17.5	14	14	14	14	14	13	13	13	13	13
18.0	15	15	15	15	15	14	14	14	14	14
18.5	17	16	16	16	16	15	15	15	15	15
19.0	18	17	17	17	17	16	16	16	16	16
19.5	18	18	18	18	18	17	17	17	17	17
20.0	19	19	19	19	19	18	18	18	18	18
20.5	20	20	20	20	19	19	19	19	19	18
21.0	21	21	21	21	20	20	20	20	20	19
21.5	22	22	22	21	21	21	21	21	20	20
22.0	23	23	23	22	22	22	22	22	21	21
22.5	24	24	23	23	23	23	23	22	22	22
23.0	25	24	24	24	24	24	23	23	23	23
23.5	25	25	25	25	25	24	24	24	24	24
24.0	26	26	26	26	25	25	25	25	25	24
24.5	27	27	27	26	26	26	26	26	25	25
25.0	28	28	27	27	27	27	26	26	26	26
25.5	29	28	28	28	28	27	27	27	27	27
26.0	29	29	29	29	28	28	28	28	28	27
26.5	30	30	30	29	29	29	29	28	28	28
27.0	31	30	30	30	30	30	29	29	29	29
27.5	31	31	31	31	30	30	30	30	30	29
28.0	32	32	32	31	31	31	31	31	30	30
28.5	33	32	32	32	32	32	31	31	31	31
29.0	33	33	33	33	32	32	32	32	32	31
29.5	34	34	34	33	33	33	33	32	32	32
30.0	35	34	34	34	34	34	33	33	33	33
30.5	35	35	35	35	34	34	34	34	34	33
31.0	36	36	35	35	35	35	35	34	34	34
31.5	36	36	36	36	36	35	35	35	35	35
32.0	37	37	37	36	36	36	36	36	35	35
32.5				37	37	36	36	36	36	36
33.0							37	37	36	36
33.5										37
34.0										
34.5										
35.0										

* Circumference Value = abdomen circumference - neck circumference (in inches)

SECTION 2: PERCENT BODY FAT ESTIMATION

PERCENT BODY FAT ESTIMATION FOR MEN

Circumference Value*	Height (in)									
	75.0	75.5	76.0	76.5	77.0	77.5	78.0	78.5	79.0	79.5
13.5										
14.0										
14.5										
15.0										
15.5										
16.0										
16.5	10	10	10	10	9	9				
17.0	11	11	11	11	10	10	10	10	10	9
17.5	12	12	12	12	12	11	11	11	11	11
18.0	13	13	13	13	13	12	12	12	12	12
18.5	14	14	14	14	14	13	13	13	13	13
19.0	15	15	15	15	15	14	14	14	14	14
19.5	16	16	16	16	16	15	15	15	15	15
20.0	17	17	17	17	17	16	16	16	16	16
20.5	18	18	18	18	17	17	17	17	17	16
21.0	19	19	19	19	18	18	18	18	18	17
21.5	20	20	20	19	19	19	19	19	18	18
22.0	21	21	20	20	20	20	20	20	19	19
22.5	22	22	21	21	21	21	21	20	20	20
23.0	23	22	22	22	22	22	21	21	21	21
23.5	23	23	23	23	23	22	22	22	22	22
24.0	24	24	24	24	23	23	23	23	23	22
24.5	25	25	25	24	24	24	24	24	23	23
25.0	26	25	25	25	25	25	24	24	24	24
25.5	26	26	26	26	26	25	25	25	25	25
26.0	27	27	27	27	26	26	26	26	26	25
26.5	28	28	27	27	27	27	27	26	26	26
27.0	29	28	28	28	28	28	27	27	27	27
27.5	29	29	29	29	28	28	28	28	28	27
28.0	30	30	29	29	29	29	29	29	28	28
28.5	31	30	30	30	30	30	29	29	29	29
29.0	31	31	31	31	30	30	30	30	30	29
29.5	32	32	31	31	31	31	31	30	30	30
30.0	32	32	32	32	32	31	31	31	31	31
30.5	33	33	33	32	32	32	32	32	32	31
31.0	34	33	33	33	33	33	33	32	32	32
31.5	34	34	34	34	33	33	33	33	33	33
32.0	35	35	34	34	34	34	34	33	33	33
32.5	35	35	35	35	35	34	34	34	34	34
33.0	36	36	36	35	35	35	35	35	34	34
33.5	37	36	36	36	36	36	35	35	35	35
34.0		37	37	37	36	36	36	36	36	35
34.5					37	37	37	36	36	36
35.0								37	37	36

* Circumference Value = abdomen circumference - neck circumference (in inches)

PERCENT BODY FAT ESTIMATION FOR WOMEN

Circumference Value*	Height (in)									
	58.0	58.5	59.0	59.5	60.0	60.5	61.0	61.5	62.0	62.5
45.0	19									
45.5	20	20	19							
46.0	21	20	20	20	19					
46.5	21	21	21	20	20	20	19	19		
47.0	22	22	22	21	21	20	20	20	19	19
47.5	23	23	22	22	22	21	21	21	20	20
48.0	24	23	23	23	22	22	22	21	21	21
48.5	24	24	24	23	23	23	22	22	22	21
49.0	25	25	24	24	24	23	23	23	22	22
49.5	26	26	25	25	24	24	24	23	23	23
50.0	27	26	26	26	25	25	24	24	24	23
50.5	27	27	27	26	26	26	25	25	25	24
51.0	28	28	27	27	27	26	26	26	25	25
51.5	29	28	28	28	27	27	27	26	26	26
52.0	29	29	29	28	28	28	27	27	27	26
52.5	30	30	29	29	29	28	28	28	27	27
53.0	31	30	30	30	29	29	29	28	28	28
53.5	31	31	31	30	30	30	29	29	29	28
54.0	32	32	31	31	31	30	30	30	29	29
54.5	33	32	32	32	31	31	31	30	30	30
55.0	33	33	33	32	32	32	31	31	31	30
55.5	34	34	33	33	33	32	32	32	31	31
56.0	35	34	34	34	33	33	33	32	32	31
56.5	35	35	35	34	34	34	33	33	32	32
57.0	36	36	35	35	34	34	34	33	33	33
57.5	37	36	36	35	35	35	34	34	34	33
58.0	37	37	36	36	36	35	35	35	34	34
58.5	38	37	37	37	36	36	36	35	35	35
59.0	38	38	38	37	37	37	36	36	36	35
59.5	39	39	38	38	38	37	37	36	36	36
60.0	40	39	39	38	38	38	37	37	37	36
60.5	40	40	39	39	39	38	38	38	37	37
61.0	41	40	40	40	39	39	39	38	38	38
61.5	41	41	41	40	40	40	39	39	38	38
62.0	42	42	41	41	40	40	40	39	39	39

* Circumference Value = waist circumference + hip circumference - neck circumference (in inches)

SECTION 2: PERCENT BODY FAT ESTIMATION

PERCENT BODY FAT ESTIMATION FOR WOMEN

Circumference Value*	Height (in)									
	58.0	58.5	59.0	59.5	60.0	60.5	61.0	61.5	62.0	62.5
62.5	42	42	42	41	41	41	40	40	40	39
63.0	43	43	42	42	42	41	41	41	40	40
63.5	44	43	43	42	42	42	41	41	41	40
64.0	44	44	43	43	43	42	42	42	41	41
64.5	45	44	44	44	43	43	43	42	42	42
65.0	45	45	45	44	44	43	43	43	42	42
65.5	46	45	45	45	44	44	44	43	43	43
66.0	46	46	46	45	45	45	44	44	43	43
66.5	47	46	46	46	45	45	45	44	44	44
67.0			47	46	46	46	45	45	45	44
67.5				47	46	46	46	45	45	45
68.0					47	47	46	46	46	45
68.5							47	46	46	46
69.0								47	47	46
69.5										47
70.0										
70.5										
71.0										
71.5										
72.0										
72.5										
73.0										
73.5										
74.0										
74.5										
75.0										
75.5										
76.0										
76.5										
77.0										
77.5										
78.0										
78.5										
79.0										
79.5										

* Circumference Value = waist circumference + hip circumference - neck circumference (in inches)

PERCENT BODY FAT ESTIMATION FOR WOMEN

Circumference Value*	Height (in)									
	63.0	63.5	64.0	64.5	65.0	65.5	66.0	66.5	67.0	67.5
45.0										
45.5										
46.0										
46.5										
47.0										
47.5	19	19								
48.0	20	20	20	19						
48.5	21	21	20	20	20	19				
49.0	22	21	21	21	20	20	20	19	19	
49.5	22	22	22	21	21	21	20	20	20	19
50.0	23	23	22	22	22	21	21	21	21	20
50.5	24	23	23	23	23	22	22	22	21	21
51.0	25	24	24	24	23	23	23	22	22	22
51.5	25	25	25	24	24	24	23	23	23	22
52.0	26	26	25	25	25	24	24	24	23	23
52.5	27	26	26	26	25	25	25	24	24	24
53.0	27	27	27	26	26	26	25	25	25	24
53.5	28	28	27	27	27	26	26	26	25	25
54.0	29	28	28	28	27	27	27	26	26	26
54.5	29	29	29	28	28	28	27	27	27	26
55.0	30	30	29	29	29	28	28	28	27	27
55.5	31	30	30	30	29	29	29	28	28	28
56.0	31	31	30	30	30	30	29	29	29	28
56.5	32	31	31	31	30	30	30	29	29	29
57.0	32	32	32	31	31	31	30	30	30	29
57.5	33	33	32	32	32	31	31	31	30	30
58.0	34	33	33	33	32	32	32	31	31	31
58.5	34	34	34	33	33	33	32	32	32	31
59.0	35	35	34	34	34	33	33	33	32	32
59.5	35	35	35	34	34	34	33	33	33	33
60.0	36	36	35	35	35	34	34	34	33	33
60.5	37	36	36	36	35	35	35	34	34	34
61.0	37	37	37	36	36	36	35	35	35	34
61.5	38	37	37	37	36	36	36	36	35	35
62.0	38	38	38	37	37	37	36	36	36	35

* Circumference Value = waist circumference + hip circumference - neck circumference (in inches)

PERCENT BODY FAT ESTIMATION FOR WOMEN

Circumference Value*	Height (in)									
	63.0	63.5	64.0	64.5	65.0	65.5	66.0	66.5	67.0	67.5
62.5	39	39	38	38	38	37	37	37	36	36
63.0	40	39	39	39	38	38	38	37	37	37
63.5	40	40	39	39	39	38	38	38	37	37
64.0	41	40	40	40	39	39	39	38	38	38
64.5	41	41	41	40	40	40	39	39	39	30
65.0	42	41	41	41	40	40	40	39	39	39
65.5	42	42	42	41	41	41	40	40	40	39
66.0	43	42	42	42	41	41	41	41	40	40
66.5	43	43	43	42	42	42	41	41	41	40
67.0	44	44	43	43	43	42	42	42	41	41
67.5	44	44	44	43	43	43	42	42	42	41
68.0	45	45	44	44	44	43	43	43	42	42
68.5	45	45	45	44	44	44	43	43	43	43
69.0	46	46	45	45	45	44	44	44	43	43
69.5	46	46	46	45	45	45	44	44	44	44
70.0	47	47	46	46	46	45	45	45	44	44
70.5			47	46	46	46	46	45	45	45
71.0				47	47	46	46	46	45	45
71.5						47	47	46	46	46
72.0							47	47	46	46
72.5									47	47
73.0										
73.5										
74.0										
74.5										
75.0										
75.5										
76.0										
76.5										
77.0										
77.5										
78.0										
78.5										
79.0										
79.5										

* Circumference Value = waist circumference + hip circumference - neck circumference (in inches)

PERCENT BODY FAT ESTIMATION FOR WOMEN

Circumference Value*	Height (in)									
	68.0	68.5	69.0	69.5	70.0	70.5	71.0	71.5	72.0	72.5
45.0										
45.5										
46.0										
46.5										
47.0										
47.5										
48.0										
48.5										
49.0										
49.5	19									
50.0	20	20	19							
50.5	21	20	20	20	19	19				
51.0	21	21	21	20	20	20	19	19		
51.5	22	22	21	21	21	20	20	20	20	19
52.0	23	22	22	22	21	21	21	21	20	20
52.5	23	23	23	22	22	22	22	21	21	21
53.0	24	24	23	23	23	22	22	22	22	21
53.5	25	24	24	24	23	23	23	23	22	22
54.0	25	25	25	24	24	24	24	23	23	23
54.5	26	26	25	25	25	24	24	24	24	23
55.0	27	26	26	26	25	25	25	25	24	24
55.5	27	27	27	26	26	26	25	25	25	25
56.0	28	28	27	27	27	26	26	26	25	25
56.5	29	28	28	28	27	27	27	26	26	26
57.0	29	29	29	28	28	28	27	27	27	26
57.5	30	29	29	29	29	28	28	28	27	27
58.0	30	30	30	29	29	29	29	28	28	28
58.5	31	31	30	30	30	29	29	29	29	28
59.0	32	31	31	31	30	30	30	29	29	29
59.5	32	32	32	31	31	31	30	30	30	29
60.0	33	32	32	32	32	31	31	31	30	30
60.5	33	33	33	32	32	32	32	31	31	31
61.0	34	34	33	33	33	32	32	32	32	31
61.5	35	34	34	34	33	33	33	32	32	32
62.0	35	35	35	34	34	34	33	33	33	32

* Circumference Value = waist circumference + hip circumference - neck circumference (in inches)

PERCENT BODY FAT ESTIMATION FOR WOMEN

Circumference Value*	Height (in)									
	68.0	68.5	69.0	69.5	70.0	70.5	71.0	71.5	72.0	72.5
62.5	36	35	35	35	34	34	34	34	33	33
63.0	36	36	36	35	35	35	34	34	34	34
63.5	37	37	36	36	36	35	35	35	34	34
64.0	37	37	37	36	36	36	36	35	35	35
64.5	38	38	37	37	37	36	36	36	36	35
65.0	38	38	38	38	37	37	37	36	36	36
65.5	39	39	38	38	38	37	37	37	37	36
66.0	40	39	39	39	38	38	38	37	37	37
66.5	40	40	39	39	39	39	38	38	38	37
67.0	41	40	40	40	39	39	39	39	38	38
67.5	41	41	41	40	40	40	39	39	39	38
68.0	42	41	41	41	40	40	40	40	39	39
68.5	42	42	42	41	41	41	40	40	40	39
69.0	43	42	42	42	41	41	41	41	40	40
69.5	43	43	43	42	42	42	41	41	41	41
70.0	44	43	43	43	43	42	42	42	41	41
70.5	44	44	44	43	43	43	42	42	42	42
71.0	45	44	44	44	44	43	43	43	42	42
71.5	45	45	45	44	44	44	43	43	43	43
72.0	46	45	45	45	45	44	44	44	43	43
72.5	46	46	46	45	45	45	44	44	44	44
73.0	47	46	46	46	45	45	45	45	44	44
73.5		47	47	46	46	46	45	45	45	44
74.0				47	46	46	46	46	45	45
74.5					47	47	46	46	46	45
75.0							47	46	46	46
75.5								47	47	46
76.0										47
76.5										
77.0										
77.5										
78.0										
78.5										
79.0										
79.5										

* Circumference Value = waist circumference + hip circumference - neck circumference (in inches)

PERCENT BODY FAT ESTIMATION FOR WOMEN

Circumference Value*	Height (in)									
	73.0	73.5	74.0	74.5	75.0	75.5	76.0	76.5	77.0	77.5
45.0										
45.5										
46.0										
46.5										
47.0										
47.5										
48.0										
48.5										
49.0										
49.5										
50.0										
50.5										
51.0										
51.5										
52.0	20	19	19							
52.5	20	20	20	19	19					
53.0	21	21	20	20	20	20	19	19		
53.5	22	21	21	21	21	20	20	20	19	19
54.0	22	22	22	21	21	21	21	20	20	20
54.5	23	23	22	22	22	22	21	21	21	20
55.0	24	23	23	23	22	22	22	22	21	21
55.5	24	24	24	23	23	23	23	22	22	22
56.0	25	25	24	24	24	23	23	23	23	22
56.5	26	25	25	25	24	24	24	24	23	23
57.0	26	26	26	25	25	25	24	24	24	24
57.5	27	26	26	26	26	25	25	25	25	24
58.0	27	27	27	27	26	26	26	25	25	25
58.5	28	28	27	27	27	27	26	26	26	25
59.0	29	28	28	28	27	27	27	27	26	26
59.5	29	29	29	28	28	28	27	27	27	27
60.0	30	30	29	29	29	28	28	28	28	27
60.5	30	30	30	30	29	29	29	28	28	28
61.0	31	31	30	30	30	30	29	29	29	28
61.5	32	31	31	31	30	30	30	30	29	29
62.0	32	32	32	31	31	31	30	30	30	30

* Circumference Value = waist circumference + hip circumference - neck circumference (in inches)

PERCENT BODY FAT ESTIMATION FOR WOMEN

Circumference Value*	Height (in)									
	73.0	73.5	74.0	74.5	75.0	75.5	76.0	76.5	77.0	77.5
62.5	33	32	32	32	32	31	31	31	30	30
63.0	33	33	33	32	32	32	32	31	31	31
63.5	34	34	33	33	33	32	32	32	32	31
64.0	34	34	34	34	33	33	33	32	32	32
64.5	35	35	34	34	34	33	33	33	33	32
65.0	35	35	35	35	34	34	34	33	33	33
65.5	36	36	35	35	35	35	34	34	34	33
66.0	37	36	36	36	35	35	35	35	34	34
66.5	37	37	37	36	36	36	35	35	35	35
67.0	38	37	37	37	36	36	36	36	35	35
67.5	38	38	38	37	37	37	36	36	36	36
68.0	39	38	38	38	38	37	37	37	36	36
68.5	39	39	39	38	38	38	37	37	37	37
69.0	40	39	39	39	39	38	38	38	37	37
69.5	40	40	40	39	39	39	39	38	38	38
70.0	41	40	40	40	40	39	39	39	38	38
70.5	41	41	41	40	40	40	40	39	39	39
71.0	42	41	41	41	41	40	40	40	39	39
71.5	42	42	42	41	41	41	41	40	40	40
72.0	43	42	42	42	42	41	41	41	40	40
72.5	43	43	43	42	42	42	42	41	41	41
73.0	44	43	43	43	43	42	42	42	41	41
73.5	44	44	44	43	43	43	42	42	42	42
74.0	45	44	44	44	44	43	43	43	42	42
74.5	45	45	45	44	44	44	43	43	43	43
75.0	46	45	45	45	44	44	44	44	43	43
75.5	46	46	46	45	45	45	44	44	44	44
76.0	47	46	46	46	45	45	45	45	44	44
76.5		47	46	46	46	46	45	45	45	44
77.0			47	47	46	46	46	45	45	45
77.5					47	47	46	46	46	45
78.0						47	47	46	46	46
78.5								47	47	46
79.0										47
79.5										

* Circumference Value = waist circumference + hip circumference - neck circumference (in inches)

APPENDIX J

PHYSICAL FITNESS TEST (PFT) ENHANCEMENT TRAINING

1. Background. The PFT is an excellent instrument for measuring a Marine's general fitness level, as it relates to upper body strength and cardio-vascular endurance. Although the PFT is not specifically designed to evaluate a Marine's level of combat fitness, it does provide a baseline for potential success in meeting the physical demands of combat. This Appendix provides an abridged guide for PFT training that both reviews and applies the general principles of strength and conditioning for each of the PFT events.

2. General Strength and Conditioning Principles

a. Specificity. The main benefit of specificity is specific training develops neuromuscular efficiency. Movement that is more efficient means more repetitions or faster times. This principle states that to become good at a particular movement, you must practice that particular movement. For example, a pistol shooter does not become good at rapid fire by shooting continual precision strings of fire. Specificity states the shooter would be better served by specific training because specific training yields specific gains.

(1) Pullup/Flexed Arm Hang. There are many exercises and machines that can enhance upper body strength, but there is no better exercise for improving pullup or the flexed-arm hang performance than the act of doing or practicing them.

(2) Crunch. The various abdominal strengthening devices (machines) and routines that focus on the mid- to lower abdominal muscles will improve crunch performance in most cases. However, doing or practicing crunches is the best method for improving crunch performance.

(3) 3.0 Mile Run. Exercise machines such as stair climbers, stationary bicycles, elliptical cross trainers, and rowing machines can enhance a Marine's cardiovascular endurance. They cannot, however replace running as the activity that will best improve running performance.

b. Overload. The acronym "SAID" stands for "Specific Adaptations to Imposed Demands," and describes what happens when the muscular and cardiovascular systems are overloaded or "pushed." When a Marine does an activity to a point of fatigue, muscle systems that have been exerted will become stronger and better able to handle the increased exercise stimulus. Proper rest and nutrition are important cofactors in ensuring the benefits of overloading are realized. It is important for the Marine and the commander to understand that improper overloading could have a detrimental effect on the desired result.

(1) Pullup/Flexed-Arm Hang. Routines that require completing set(s) of pullups to the point of exhaustion are worthwhile. Form and slowed execution of the pullup should be maintained in order to facilitate momentary muscular fatigue. Executing pullups under weight (e.g., with a pack or flak jacket on) and varying the width of the grip (e.g., narrow to wide grip) are excellent training methods. Training for pullups in this manner will help ensure that a Marine will be able to do proper pullups in an all-out effort on test day.

(2) Crunch. Applying the principle of overload in preparing for the crunch requires doing two sets of crunches until momentary muscular fatigue. Tracking the amount of "complete" crunches a Marine can do until the point of fatigue, in a certain time period, will provide excellent feedback on the effectiveness of the overload principle. A second way to train for the crunch event would be for a Marine to do as many crunches as he can, without stopping, until complete muscle failure. Focusing on using the abdominal muscles (keeping them tight throughout the exercise), and executing them in slow motion is an excellent way to intensify the training. Track the number of proper repetitions executed to determine effectiveness.

(3) 3.0 Mile Run. Interval training done once or twice a week can enhance a Marine's running performance. Interval training sessions take less time, but require a higher level of intensity to run the shorter distances. The 400-meter and 800-meter courses are the best distances for increasing performance for the 3.0 mile run. If the distance interval training were longer than 800 meters, the Marine would not be able to run at the targeted speed. If the distance is shorter than 400 meters, the Marine cannot reach the targeted level of intensity needed for endurance training.

(a) To calculate the time an individual Marine should run the 400/800-meter distances during an interval training session is as follows:

1 Determine the most recent 3.0 mile run time;

2 Divide that time by 3. This gives the mile-split time for the run;

3 Divide the mile-split time by 2. Subtract that time by 10-20 seconds. This gives the time the 800-meter intervals should be run.

4 To determine the 400-meter interval time, divide the mile-split time by 4. Then subtract 5-10 seconds. This is the time the 400-meter intervals should be run.

(b) Calculating rest in between each running exercise is very important. Complete recovery in between the intervals lessens the benefit of overload, but not enough rest is also detrimental. Incomplete recovery in between intervals should be the goal. The work/rest ratio for 800-meter intervals should be 1:1. In other words, active rest in the form of light jogging or walking (not stopping) should be equal to the time that was run. For example if a 4:00 minute interval time was run, then rest should be 4:00 minutes. The work/rest ratio for 400-meter intervals should be 1:2. Using the same type of rest as in 800-meter interval session, a 1 min 40 second 400-meter interval would be followed by a 3 min 20 second rest period.

1 Example: A Marine wants to do interval sessions to improve the 22:00 he ran in the last PFT.

a Divide 22:00 by 3 (7:20);

b Divide 7:20 by 2 to determine the pace for an 800-meter split time for a 7:20 mile pace. (3:40);

c Subtract 10-20 seconds to determine the pace for 800-meter intervals. (3:20-3:40);

d The rest between these 800 repeats should be 3:20-3:40.

e Determine the pace for a 400-meter interval by dividing 7:20 by 4 to determine the pace for a 400-meter split time for 7:20 mile pace (1:50). Subtract 5-10 seconds to determine pace for 400-meter intervals. (1:40-1:45);

f The rest between these 400 repeats should be 3:20-3:30.

2 A sample (beginner's) interval workout:

- Daily 16 Warmup
- 800 meter jog
- 800-rest-800-rest-400-rest-400
- 800 meter jog
- Daily 16 Cool-Down

(c) The total mileage for this workout is 2.5 miles with 1.5 miles being speed-work. Many Marines and recreational runners do intervals that are much too fast for their goal time. Keeping the predetermined goal pace should be the goal for this workout. A Marine should not feel totally exhausted following this training session. As a Marine gets more comfortable with this training, more intervals can be added.

(d) Interval training should be combined with other types of running, during the weekly training schedule, for the training to be effective. Longer distance runs along with tempo running (faster running but not at race pace), combined with interval running will improve most Marines' 3.0 mile run times.

3. Summary

a. Specificity and overload are two of the most important strength and conditioning principles. Other principles include:

(1) Regularity - Strength and cardiovascular training should be done 3-4 times a week.

(2) Variety - Varying training regimes periodically prevents boredom.

(3) Recovery - Rest is physiologically necessary for growth and repair.

(4) Balance - A good training plan should exercise all major muscle groups, and include warmup and cool-down phases.

(5) Progression - Ensures that systems are continually challenged.

b. There are many ways in which Marines can improve their performance on the PFT. By applying the strength and conditioning principles listed above, training can be enhanced to improve not only PFT performance but increase a Marines' physical readiness for combat.

APPENDIX K

LIGHT/LIMITED DUTY ASSIGNMENT PROCESS

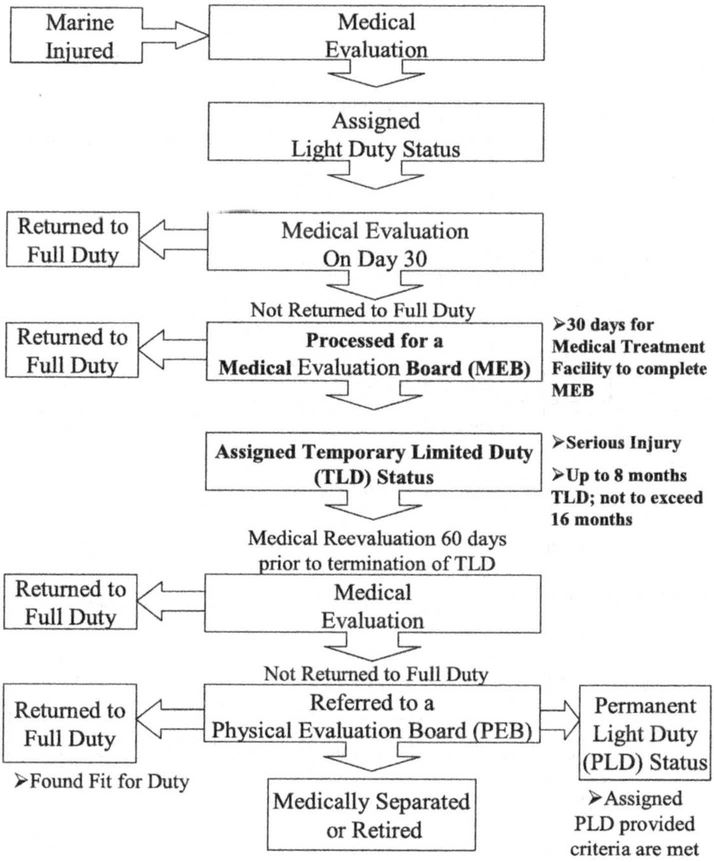

NOTE:

- Marines on light or limited duty (TLD & PLD) are required to maintain Height/Weight or Body Fat Standards regardless of duty status. Marines not maintaining standards can be assigned to the Body Composition Program.
- Marines assigned to light or limited duty (TLD & PLD) who are medically restricted from completing all three events of a PFT will be screened for participation in a Partial PFT – 2 of the 3 events.

APPENDIX L

PREGNANCY PFT AND BODY COMPOSITION PROGRAM FLOWCHART

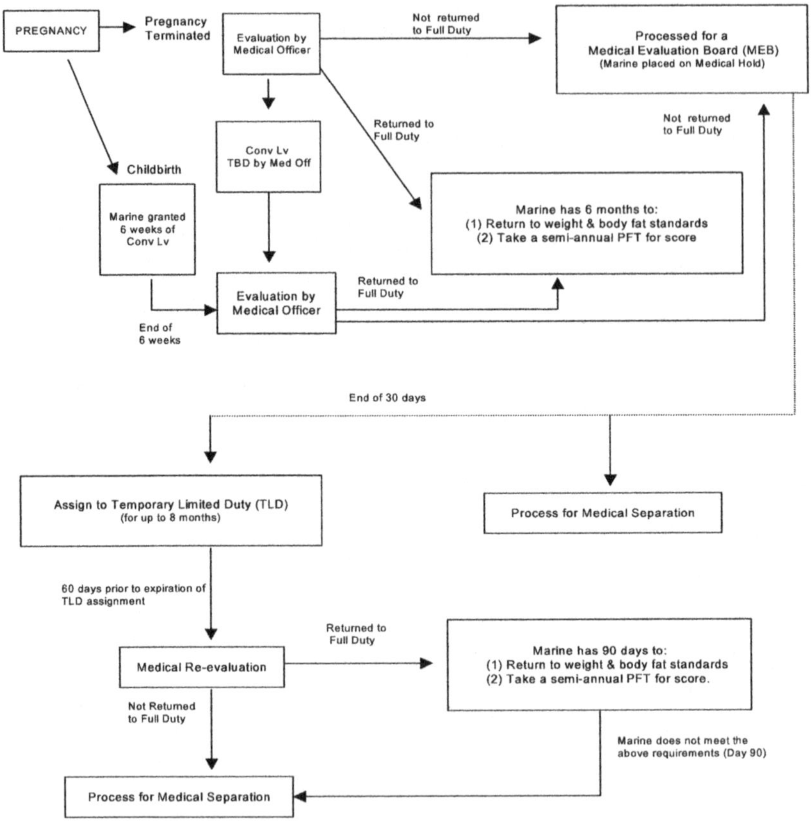

NOTE: Marines assigned to a Body Composition Program (BCP) PRIOR to pregnancy will go into an "inactive status". Once returned to full duty, Marine will return to an active status and will have 6 months from that day to return to weight & body fat standards, regardless on time spent on Program previously.

Made in the USA
Monee, IL
07 November 2026

68027073R00134